# A SHORT HISTORY

# OF

# ARYAN MEDICAL SCIENCE

# A SHORT HISTORY

# OF

# ARYAN MEDICAL SCIENCE

BY

**BHAGVAT SINH JEE**

GYAN

*Published by*

GYAN

Gyan Publishing House
5, Ansari Road
Daryaganj, New Delhi-110002
Phone: 011-47034999, 9811692060
E-mail: books@gyanbooks.com

*Distribution Network*
gyanbooks.com
India, USA, Canada, UK, Australia



ISBN: 978-93-6433-227-9 (PB)
First Published, 1966

2nd Impression 2024

*Printed at:* Gyan Press, Delhi.



**A SHORT HISTORY OF ARYAN MEDICAL SCIENCE**
Author: BHAGVAT SINH JEE

# A SHORT HISTORY

OF

# ARYAN MEDICAL SCIENCE

# A SHORT HISTORY

OF

# ARYAN MEDICAL SCIENCE

BY

H.H. Sir BHAGVAT SINH JEE, K.C.I.E.

M.D., D.C.L., LL.D., F.R.C.P.E.

THAKORE SAHEB OF GONDAL

WITH TEN PLATES

London

MACMILLAN AND CO., Ltd.

NEW YORK: THE MACMILLAN COMPANY

1896

TO

PROFESSOR SIR WILLIAM TURNER

M.B., LL.D., D.C.L., D.SC., F.R.S., ETC.

IN CHERISHED REMEMBRANCE OF

MANY HAPPY HOURS

This small Volume is inscribed

BY

HIS GRATEFUL PUPIL AND ADMIRER

# PREFACE.

THE literature of ARYAN MEDICINE is so vast and copious that it is impossible to do adequate justice to its history within the compass of a small manual. The science is acknowledged on all hands to be of great antiquity, and contains a mine of information not to be altogether despised by the students of medicine of our day. Should it be approached in a spirit of fairness and inquiry, possibly it might disclose the germs of not a few of the marvellous discoveries in the realm of Medicine of which the present century is justly proud, and afford a diligent scholar ample scope and materials for comparison between the old and the new systems, with a view to supply the deficiencies of the one or the other for the benefit of mankind. An elaborate and complete history of Hindoo Medical Science is a great desideratum. If this humble attempt, however imperfect and fragmentary, should induce any of my readers to set himself seriously to the task of compiling from original sources a more readable and comprehensive history, I shall consider my labour to have been well spent.

B. S. J.

GONDAL, 1895.

# CONTENTS.


| CHAP. | | PAGE |
|---|---|---|
| I. | Early Civilisation of the Hindoos, . . . . | 11 |
| II. | Ancient Writers on Hindoo Medicine, . . | 24 |
| III. | The Hindoo Theory of Creation, . . . | 40 |
| IV. | Hindoo Practice during Period of Nubility, | 45 |
| V. | Principles of Hygiene as understood by the Hindoos, . . . . . . . . . . | 56 |
| VI. | Theory of Indian Medicine, . . . . | 85 |
| VII. | Indian Materia Medica, . . . . . . | 101 |
| VIII. | Hindoo Writers on Ætiology, Diagnosis and Treatment, . . . . . . . . . . | 151 |
| IX. | Qualities of a Physician and his Prognosis, . | 161 |
| X. | Indian Surgery—its Rise and Fall, . . . | 176 |
| XI. | Vicissitudes of Indian Medicine and Surgery, . | 187 |
| XII. | Concluding Remarks, . . . . . . . | 201 |
| | Bibliography, . . . . . . . . | 207 |
| | Index and Glossary, . . . . . . . | 217 |

# LIST OF ILLUSTRATIONS.

PLATES I.-VI.
PHARMACEUTICAL APPARATUS.

PLATES VII.-X.
SURGICAL APPLIANCES AND INSTRUMENTS.

# SHORT HISTORY

OF

# ARYAN MEDICAL SCIENCE.

## CHAPTER I.

### EARLY CIVILISATION OF THE HINDOOS.

THE history of Aryan Medicine forms an inseparable chapter of the history of Aryan civilisation. The word 'Aryan' is here used in what the Hindoos believe to be the original and only proper sense. It has been customary of late years to give it a much wider meaning, so as to make it denote the supposed original people, from whom, according to the Western Ethnologists, Celts and Teutons, Italians and Greeks, Persians and Hindoos, are all descended. The similarities which modern science has discovered between such outlying members of a supposed original stock as the Celts and the Hindoos, have led certain scholars to believe that the ancestors of these nations were first living together in the Caucasus, but afterwards separated, the Hindoos migrating into India, where they settled with their families after conquering the aboriginal tribes. This theory is European in its conception, and is not accepted by the Indians in general, who call themselves autochthonous. The Indian *Savants* adduce internal and external evidence to show that, far from being outsiders, it was possible for the early Hindoos to have sent colonies beyond the frontiers. It is no part of our present purpose to attempt to compose this controversy. It is enough to note at starting that throughout these pages the term 'Aryan' is used to designate the 'Hindoos,' to whom alone it is applied both in common parlance and in their sacred books.

The Hindoos call their country 'ARYAVARTA,' or the abode of the Aryans. Manu, the ancient law-giver, applies the name to the tract of land between the Himalaya and the Vindhya ranges, from the eastern to the western sea; and teaches that the Brahmans born within that tract are suitable teachers of the several usages of men. Latterly the whole country from the Himalayas to Cape Comorin, and from the Irawady and the Bay of Bengal to the Indus and the Western sea, came to be recognised as ARYAVARTA. It is a beautiful country with natural boundaries. It enjoys the six seasons of the year, and the position of its mountains and its seas gives it a variety of climate, inasmuch as it possesses the hottest, coolest, and the most temperate places of resort. The country was a cradle of learning for the whole world, and history bears witness to the fact that many a nation that now walks with its head erect would have been nowhere had it not borrowed considerably from the intellectual storehouse of the ancient Hindoos. This country was at the pinnacle of glory when other nations were either not in existence or were wallowing in crass ignorance. Most of the sciences, which the present century boasts of so much, were not unknown to the ancient Hindoos; and one has but to look into their writings to see whether the truths propounded by them some thousands of years ago do not still endure in their natural freshness.

The Hindoos were the first to cultivate Astronomical Science (*Jyotisha*). All modern astronomers admit the great antiquity of their observations. Cassini, Bailly, and Playfair have stated that observations taken by Hindoo Astronomers upwards of 3000 years before Christ are still extant, and prove a considerable degree of progress already made at that period. The ancient Hindoos fixed the Calendar, observed and predicted the eclipses, and were acquainted with the phases of the moon and the motions of the several planets. According to Mr Colebrooke, they were more correct than Ptolemy in their notions regarding the precession of the equinoxes.

In Mathematics (*Ganita*) the Hindoos had attained a high degree of proficiency. They invented the decimal system, the differential, integral and infinitesimal calculi. The world owes to them the invention of numerical symbols. They also discovered Geometry (*Bhoomiti*) and Trigonometry (*Triconamiti*), in both which sciences they made great advances. Most of the credit given to Pythagoras for the discovery of mathematical truths properly belongs to the ancient Hindoos.

Their knowledge of Chemistry was not meagre. They were familiar with the preparation of sulphuric, nitric, and muriatic acids; the oxides of copper, iron, lead, tin, and zinc; as well as many chlorides, nitrates, sulphates, and carbonates.

The sage Panini was the first to teach the formative principles of words, and his system of Grammar, called *Ashtadhyayi*—the first in the world—is the admiration of Western and Eastern scholars. Lexicography was known to the Aryans long before its acquaintance was made by any other nation in the world. In the Vedic Literature it is treated under the head of *Nighantu*.

Music appears to have been cultivated to the highest pitch of perfection by the Aryans, who were the first to invent the *Gamut*. Their music is systematic and refined.

India is the home of architectural beauty. Domes, cupolas, minarets, and many ingenious works of architecture which have stood the tempest of time, testify to this fact in silent eloquence; and the ancient Greeks, who are praised for their skill in this particular art, owed not a little to the Hindoos. Dr W. W. Hunter supposes that Alexander the Great left artists in India to copy the Indian style of architecture, who imported it into their mother country.

*Dhanur-Veda* is an old science which treats

of the art of war, and mentions different kinds of weapons classified under four heads: (1) *Mukta* (missive), as the discus, etc. (2) *Amukta* (non-missive), as the sword, etc. (3) *Muktamukta* (both missive and non-missive), as the javelin, etc. (4) *Yantra-mukta* (machine-projectile), as the arrow, etc. The army consisted of infantry, cavalry, car-fighters and warriors fighting on elephants. They were known by the name of *Padati*, *Ashvarudha*, *Ratharudha*, and *Gajarudha* respectively. The Hindoos have had from a primeval period a fighting class called the *Kshatriyas*.

Hindoo Law is as old as their religion. Manu is the oldest of Hindoo writers on Law; and his book of Institutes still forms the basis of the Hindoo social fabric. It is an important record of Hindoo society at least three thousand years old. Other writers on Law, like Yajnavalkya, Parashara, etc., are also held in great reverence, and are still quoted as high authorities in deciding subtle points of dispute.

India out-distances all the countries in the world in the domain of Philosophy. There are six systems of Indian Philosophy, called *Darshanas*, or 'Mirrors of Knowledge.' These are

*Nyaya* or logical, *Sankhya* or discriminative, *Vaisheshika* or atomic, *Yoga* or contemplative, *Mimansa* or ritual, and *Vedanta* or the end of knowledge. The aim and object of these schools is to solve the problem of Creation. The Hindoos have a passion for philosophy, and have given their best energies to the better understanding of the subject. They were the first nation to distinguish between matter and spirit. While the world at large has been busy confining its attention to dead matter and its properties, the Hindoo from the very dawn of history has devoted himself staunchly to the study of the spirit. Professor Max Müller justly observes that the Indian Aryan lives this life with a full consciousness of his being a temporary sojourner, who has no permanent interest whatever in the things of this world. Being given to spiritual pursuits rather than to earthly comforts, he is by nature better fitted to solve the problem of existence which puzzles many a thinker and metaphysician of our age.

All these branches of learning take their origin from the book of religion called the *Veda* (Knowledge), from Sanskrit *vid* (Latin, *videre*), to know. This the Indians believe to be the

Knowledge of the Universal Spirit, as distinguished from the knowledge of an individual mortal. The Aryans believe that the creation has a maker, who is eternal and is without a cause, and who, as He has evolved the Universe out of His inner consciousness, is a Knowing Being, and, being Knowing and Eternal, is all Happiness which knows no diminution. The Veda is supposed to be His revealed knowledge. Knowledge, they believe, is acquired and not created. If knowledge could be created, instruction, they argue, would, as a rule, become futile. From time immemorial it is being handed down from father to son, from preceptor to disciple. The Indians therefore trace all knowledge under the sun from the Supreme High, who is the fountain-head of learning, *Ishanas sarvavidyanam* (*Yajur Ved*), *i.e.*, Lord of all kinds of knowledge, the source from which all knowledge flows. So they will never accept a statement unless it is supported by the testimony of what has been revealed to them in their Scriptures, or by the testimony of bygone ages. Their line of investigation thus essentially differs from that followed by the modern investigators, who are solely guided by their intelligence in establishing a truth, which must remain under trial until Science in its progressive course has reached its goal.

The *Vedas* are four in number, viz. :—*Rig Veda*, *Yajur Veda*, *Sama Veda*, and *Atharva Veda*. We will not pause to discuss the various points by which the Brahmans, who are the custodians of the sacred lore of India, try to establish the eternity of their inspired writings. Suffice it to say, that in computing time by regular divisions and assigning dates to events of antiquity, the ideas of the Eastern and the Western people widely differ. Some Western authorities assert that man did not exist on the surface of the earth prior to B.C. 3000 ; while in Indian cosmogony deeds are recorded of persons said to have flourished in the previous *Yugas* or cycles of time thus divided :—

*Krita Yuga* lasted for 1,728,000 years.
*Treta Yuga* lasted for 1,296,000 years.
*Dvapara Yuga* lasted for 864,000 years.
*Kali Yuga* will last for 432,000 years.

The *Kali Yuga* is the present age of the world, and is said to have begun on Friday, the 18th February, B.C. 3102. These cycles go on revolving like the wheel on its axle, and bear some resemblance to the golden, silver, brazen and iron ages of the Greeks. Even the European chronologists, who, according to the Hindoos, are always disposed to modernise events, admit that the Vedas must have been composed about 4000 years ago. It has not been shown that any specific book was extant at a time when the Vedas were not in existence. This, at any rate, makes the Vedas older than any other writing on the surface of the earth. In the works of Manu and Panini who, according to the Western Orientalists flourished in B.C. 800 (Wilson) and B.C. 600 (Goldstücker) respectively, the Vedas are described as eternal (*Anadi*). Thus in point of antiquity the Vedas stand pre-eminently the first. Some European scholars, attempting to translate portions of the Hindoo Scriptures by the help of grammar and dictionary, but failing to grasp the real meaning, have—and no wonder—succeeded in belittling the sublime ideas therein contained. For the true interpretation of an extremely old and esoteric work, taught and learnt by the initiates only, must be acquired from those who have from generations past studied it systematically with the help of the key they possess. Still, according to the showing of these very scholars, the civilisation of the Vedic period can compare favourably with the civilisation of our modern times. The Vedic Aryan cultivated his land (vide *Rig Veda*, iii. 8, which says: "Let the bullocks carry the load and the cultivators till the ground; let the plough cut up the earth well"), and lived in neat and handsome mansions (vide *Rig Veda*, i. 2: "O Earth, give us large and habitable palaces!"). He wore neck ornaments and ear rings. The patriarch considered it his sacred duty to be a warrior, and he attended military classes for his education. He was protected by his armour (*Rig Veda*, i.); musicians were employed to chant hymns. Elephants were trained, and horses were gorgeously caparisoned. Artisans were liberally patronised for their manufactures. The *Yajur Veda* mentions weavers, sculptors, carpenters and other artisans, besides almost all the articles of manufacture generally used by a refined society. Women were well dressed, and held a high social position. The people had advanced in political condition. The *Krishna Yajur Veda* (i. 2–9) mentions kings, queens, commanders-in-chief, coach-drivers, magistrates (*puradhyaksha*), village officers, treasurers, revenue collectors (*bhagadugh*), and other accessories of an established government. Honesty in mercantile transactions is referred to in *Rig Veda*, iii. 6, wherein also are mentioned stone-built cities. Other references might be given representing the Vedic Aryan as well versed in war and politics, bright, clever, merciful, righteous and devoted to the protection of his family. Some Western scholars have hazarded an opinion that the Vedic Aryans were not acquainted with the art of writing. But this statement is not supported by evidence. On the contrary, we meet both in the *Rig Veda* and the *Yajur Veda* with such expressions as *likhitam* (written), *khurbhuj* (pen), *vacham pashyan* (seeing the words, *i.e.*, reading) and so forth. The religion and philosophy inculcated in the *Veda* are acknowledged to be of the sublimest character. All this unmistakably proves that the Aryans were the most enlightened race in the dawn of history. Such a state of civilisation, which exercises a potent influence on Indian society even to this date, could not have been attained in a day. It must have required a long course of training, and must take the nation back to the remotest antiquity. When the state of civilisation was so perfect, and when all sorts of useful sciences were regularly studied, there should be no wonder if the science of Medicine too received its share of attention. This Science forms part of the Vedas, and is called "*Ayur Veda,*" or the "Science of Life." It is based on *Rig Veda* in so far as it relates to the knowledge of medicine, while the surgery it treats of seems to have derived its origin from the *Atharva Veda.* Though an *Upa Veda* or Supplemental *Veda,* the science is considered to be co-existent with the "First Teacher," who is the "primary cause" of the whole universe. The science has passed through various vicissitudes. It will not therefore be out of place to trace its origin and development as succinctly as possible.

# CHAPTER II.

## ANCIENT WRITERS ON HINDOO MEDICINE.

AS has been stated in the preceding chapter, the Hindoos believe that, like all their other sciences, the science of Medicine has been revealed to them. YAJUR VEDA, Chapter V., speaks of God as "*Prathamo Daivyo bhishak,*" *i.e.*, the "first Divine Physician," "who drives away all diseases." Another Vedic verse addresses Him as "*Bhishaktamam tva bhishajam shrunomi,*" * which means, "I hear Thou art the best among physicians." Elsewhere He is styled "the depositary of all sciences, and physician for all worldly ills." BRAHMA, or the first member of the Hindoo triad, was the first to propound the Healing Art. He composed the AYUR VEDA, consisting of one hundred sections (*adhyayas*) of one hundred stanzas (*shlokas*) each. This sacred medical work treats of the subject of life, describes

* *Rig Veda*, ii. 7, 16.

the conditions tending to prolong or shorten life, dwells on the nature of diseases, their causes and methods of treatment. It is the oldest medical book of the Hindoos, and is divided into eight parts or *tantras*. These are:—

1. SHALYA — Surgery. This includes the methods of removing foreign bodies, of using surgical instruments, of applying bandages, and of treating various surgical diseases.
2. SHALAKYA—Treatment of diseases of parts situated above the clavicles, such as the diseases of the eyes, nose, mouth, ears, etc.
3. KAYA-CHIKITSA—General diseases affecting the whole body, such as fever, diabetes, etc.
4. BHOOT-VIDYA — Demoniacal diseases. This chapter describes the means of restoring by prayers, offerings, medicines, etc., deranged faculties of the mind supposed to be produced by demoniacal possession.
5. KAUMARA-BHRITYA—Management of children —comprising the treatment of infants and the diseases they are subject to.

6. AGADA — Antidotes for poisons — mineral, vegetable, and animal.
7. RASAYANA — Treats of medicines preserving vigour, restoring youth, improving memory, and curing and preventing diseases in general.
8. VAJIKARANA — Describes the means of increasing the virile power by giving tone to the weakened organs of generation.

BRAHMA taught Ayur Veda to DAKSHPRAJAPATI, who in turn expounded it to ASHVINI KUMARS, "the twin sons of the Sun." The twin brothers wrote important works on medicine and surgery, and were the divine physicians. Many hymns in the RIG VEDA are addressed to these twin gods, from which it appears that medicine and surgery were fully appreciated by the ancients, and held in high esteem by them. Some of the wonderful operations performed by them are also recorded. A legend in the RIG VEDA thus describes their skill: A certain sage named DADHYANCHI had learnt the science of *Brahma-vidya* from Indra under an interdiction not to teach it to any one else, the preceptor threatening to cut off his pupil's head in case of any infringement of the compact. The Ashvins, anxious to learn the science from that sage, hit upon the following plan. They, by undertaking to preserve him from the wrath of Indra, prevailed on the sage to communicate his knowledge to them. Then, with his consent, they took off his head, replaced it skilfully with that of a horse, and acquired the wished-for knowledge. Indra, when he came to know that DADHYANCHI had broken the contract, struck off the sage's equine head. The Ashvins, being exceedingly proficient in surgery, rejoined the original head that had been carefully preserved. The feat excited universal approbation. But some of the fastidious gods took exception to the mode of learning adopted by the Ashvins. The cutting off one's preceptor's head, though with the best of intentions, was denounced as an atrocious act; and as a consequence, the Ashvins were outcasted by the gods for the unpardonable sin, and refused admittance to their share in the sacrificial rites. The brothers then had recourse to a sage named Chyavana, who, though very old and decrepit, had newly married Sukanya, a young and charming daughter of king Sarayati. The physicians prescribed him an electuary, which soon freed him from his decrepitude, restored him his health, youth, and vigour, and prolonged his life (*Rig Veda*, i. 117, 13). The recipe is still known by the name of "Chyavana Avaleha." The sage, out of gratitude, promised the Ashvins to intercede in their behalf, and to secure to them the continuance of the libation of *Soma* at the sacrifices. He induced his father-in-law, king Sarayati, to perform a sacrifice. When the time for distributing the libation arrived, Chyavana offered to the Ashvins the share due to them. Indra took umbrage at this, and was going to hurl his thunderbolt at the sage's head when he found his arm suddenly paralysed. The Ashvins cured Indra of his paralysis, and by dint of their skill and knowledge soon got themselves re-admitted into caste, and obtained their usual share of sacrificial food. These physicians are also given credit for joining again the head and body of Yajna, son of Ruchi, which were severed by Rudra. In the ancient Sanskrit writings we often read of battles between the DEVATAS and ASURAS. In cases of broken legs, the surgeons used to substitute "iron-legs"—*Ayasin-jangham*—(vide *Rig Veda*, i. 116, 15), and to furnish artificial eyes in place of those plucked out (*Rig Veda*, i. 116, 16); arrows lodged in the bodies of the warriors were skilfully extracted, and their wounds promptly dressed by the army surgeons. The Ashvins are reputed to have given new teeth to Poosha, new eyes to Bhagdeva, and to have cured Chandramas of consumption. These and many other wonderful cures effected by them raised them not only in the estimation of their compeers but also of the lord INDRA, who became desirous of studying the AYUR VEDA, and learnt it from them.

Indra taught the science to his pupil ATREYA, who wrote several works bearing his name, among which might be mentioned his "Atreya Samhita," in five parts, containing 46,500 verses in all. Atreya is one of the oldest authorities on Hindoo Medicine, and several later writers have based their treatises on his work. He imparted his knowledge, among others, to AGNIVESHA, BHEDA, JATUKARNA, PARASHARA, KSHIRAPANI and HARITA, all of whom have distinguished themselves as authors of medical works that have been handed down to posterity. Agnivesha's "Nidananjana," or treatise on Diagnosis, is still admired. "Harita Samhita" is a standard book, which appears to have been dictated by Atreya in reply to Harita's questions; for each chapter ends with the words, "Said by Atreya in answer to Harita." Some are led to believe that "Atreya Samhita" and "Harita Samhita" are identical. This does not seem to be correct. For the well-known author of "Bhavaprakasha" quotes several verses from Atreya which are not found in Harita.

CHARAKA, an early medical writer, gives the origin of the healing art upon the earth as follows :—

Once upon a time some distinguished sages happened to meet on the Himalaya mountains, among them being AGASTI, ASHVALAYANA, ASITA, BADARAYANA, BALIKHYA, BHARADVAJA, CHYAVANA, DEVALA, DHAUMYA, GALAVA, GARGA, GAUTAMA, GOBHILA, HARITA, HIRANYAKSHA, JAMA DAGNI, KAMYA, KANKAYANA, KAPINJALA, KASHYAPA, KATYAYANA, KAUNDINYA, KUSHIKA, LANGAKSHI, MAITREYA, MARKANDEYA, NARADA, PARASHARA, PARIKSHAKA, PULASTYA, SANKHYA, SANKRITYA, SHAKUNEYA, SHANDILYA, SHARALOMA, SHAUNAKA, VAIJEPAYA, VAIKHANASA, VAMADEVA, VASISHTHA, VISHVAMITRA, and many others. All of them were well versed in philosophy and practised religious austerities. The subject of their conversation was the "ills that flesh is heir to." They began to complain : "Our body, which is the means of attaining the four aims of life, viz., virtue, worldly pursuits, pleasure, and liberation, is subject to diseases which emaciate and weaken it, deprive the senses of their functions, and cause extreme pain. These diseases are great impediments to our worldly affairs and bring on premature death. In the face of such enemies, how can men be happy? It is necessary therefore to find remedies for such diseases." They turned to sage Bharadvaja, and thus addressed him:—

"O sage, thou art the fittest person among us: Go thou to the thousand-eyed INDRA, who has systematically studied the Ayur Veda, and by acquiring from him the knowledge of that science free us, O sage, from the scourge of diseases."

"So be it," said the sage; who at once went to Indra and thus accosted him:—

"O Lord, I have been deputed by the parliament of sages to learn from you the remedies for the direful diseases that afflict mankind. I pray you, therefore, to teach me the Ayur Veda."

Indra was pleased with the object of his mission, and taught him the Ayur Veda in all its parts. BHARADVAJA recounted the precepts he had acquired to the other sages who had deputed him, and with the knowledge of the science they were able to live long in health and happiness.

No history of the earliest writers on Medicine in India would be complete without a mention of CHARAKA and SUSHRUTA, who are considered by the natives to be the highest authorities in all medical matters. Charaka is said to have been an incarnation of Shesha—the Serpent-god with a thousand heads—who is supposed to be the depositary of all sciences, especially of medicine. It may be parenthetically noted here that the serpent in all ages has received divine honours, and from the remotest antiquity has been held in the greatest veneration as an emblem of wisdom and immortality by the Egyptians, Greeks, and other ancient nations as well as by the Hindoos. "Serpents were sacred to Æsculapius, the Grecian god of the medical art, because they were symbols of renovation, and were believed to have the power of discovering healing herbs" (Dr Smith). The hierophants of Egypt styled themselves the "Sons of the Serpent-god," as the serpent was the emblem of wisdom and eternity. Ophite-worship was prevalent among the Jews 2000 years B.C. The fifth day of the month of Shravaṇa (which falls in the rainy season) is to this day held by the Hindoos as sacred to the serpent, which is worshipped either alive or in effigy by every mistress of a family. For it is believed that leprosy, ophthalmia, and childlessness are the punishment of those who in former lives, or in the present one, may have killed a snake, and that it is only by serpent-worship that these penalties can be averted. CHARAKA, the son of Vishudha, a learned *Muni*, flourished during the Vedic period. Some believe him to have been born at Benares 320 years B.C. He was the greatest physician of his day, and his "Charaka Samhita" is still held to be a standard work on Medicine.

SUSHRUTA, on the other hand, dilates more on Surgery than on Medicine. His work "Sushruta" is therefore held in high esteem by native *Vaidyas* as an authority on Surgery. Both the works are compendiums of the Ayur Veda. Sushruta was a son of Vishvamitra, a contemporary of Rama. With his father's permission, Sushruta and his seven brothers went to DEVODASA, king of Benares, to study Medicine. As Charaka is believed to be an incarnation of the Serpent-god, so is Devodasa believed to be an incarnation of DHANVANTARI, the divine physician, recovered from the ocean along with thirteen other *Ratnas* (gems) which had been lost in the Deluge. Dhanvantari is said to have come out of the ocean with a cup of *Amrita*, or the beverage of immortality; and he takes in India the place occupied by Æsculapius amongst the Greeks. Having learnt Ayur Veda from Devodasa or Kashiraja, as he is otherwise called, Sushruta and his companions returned home and wrote independent works on Medicine and Surgery. But Sushruta excelled them all. His work was translated into Arabic before the end of the 8th century A.C. It has been translated into Latin by Hepler and into German by Vullers. Charaka was also translated from Sanskrit into Arabic in the beginning of the 8th century, and "his name repeatedly occurs in the Latin translations of Avicenna, Razes, and Serapion" (Hunter). He was posterior to Agnivesha, for he states that he received the materials for his book from that learned sage, whose work he re-cast.

The next authority on Hindoo Medicine is VAGBHATA, who flourished about the second century before Christ. He was an inhabitant of Sindh, in Western India. In his work called "Ashtanga-hridaya," he acknowledges the assistance derived from the writings of Charaka, Sushruta, Agnivesha, Bhela, and others who had gone before him. He also wrote another work called "Ashtanga-Sangraha," on which Pandit Arunadatta wrote a commentary. Vagbhata's style is very clear and concise, and throws much light on several obscure passages in his predecessors' works. A popular couplet describes Vagbhata, Sushruta, and Atreya as the three great medical authorities for the three Yugas—Kali, Dvapara, and Krita respectively. Among the students of Hindoo Medicine the three writers are known by the name of "Vridha Trayi," or the "Old Triad."

Coming nearer to our period we meet with the name of MADHAVA or MADHAVACHARYA, who wrote several works embracing almost all branches of Hindoo learning. He was born in Kishkindha, now called Golkonda, in Southern India, and was Prime Minister to Raja Vira Bukka of Vijayanagar, in the 12th century. He was a brother of Sayana, the author of the great Commentary on the *Rig Veda*, to which work Madhava is said to have contributed. Besides the "Sarva-darshana-sangraha," or dissertation on the six schools of Hindoo philosophy, and the scholia on the four Vedas, styled "Madhava Vedartha Prakasha," the "Panchadashi" (on Vedantic philosophy), "Madhava Vritti" (on Grammar), "Madhava Nidana" (on Medicine), "Kala Madhava" (on Astronomy), "Vyavahara Madhava" (on Hindoo Law), "Achara Madhava" (on the usages of the Brahmanas), and "Shankara Digvijaya" (Life of Shankaracharya) are some of his numerous works. In his medical work our author dwells exclusively on the diagnosis of diseases. He has treated the subject so well that his authority on this branch of Medical Science is held to be indisputable. The native doctors are often heard to repeat this Sanskrit stanza :—

*Nidane Madhavas shreshthas,*
*Sutrasthane tu Vagbhatas :*
*Sharire Sushrutas proctas,*
*Charakas tu chikitsake.*

It means : Madhava is unrivalled in Diagnosis, Vagbhata in Principles and Practice of Medicine, Sushruta in Surgery, and Charaka in Therapeutics. In his old age Madhava became an ascetic, and assumed the name of VIDYARANYA ('forest of learning').

The next celebrated writer on Hindoo medicine is BHAVA MISHRA, author of the "Bhava Prakasha." This physician lived in 1550 A.C., and was considered to be the best scholar of his time in Madra Desha (in North-West of India), "a jewel of physicians and master of the Shastras." In his work he summarises the practice of all the best previous writers on Medicine. The clearness of his style and the excellence of his arrangement have thrown a flood of light on many obscure and disputed passages of the ancient writers; and his important compilation marks the last revival of Ayur Vedic literature among the Hindoos. The work is highly esteemed by native doctors in all parts of India as an invaluable treatise on Hindoo medical science. It is considered a thesaurus of useful information gleaned from the vast field of medical literature of the past. In the time of Bhava Mishra, India had commenced to come into contact with some of the European nations, notably the Portuguese, who were attracted to India by commercial pursuits. A syphilitic disease, in which hands and feet are affected, was common among the Portuguese. Bhava Mishra treats of this affection at length under the name of *Firanga Roga*, *i.e.*, Portuguese disease. The absence of a corresponding Sanskrit term, and the name (" Firanga Roga " ) given to the malady, would suggest that it was introduced into India by the Portuguese. Bhava Mishra describes three stages of the disease, namely *Bahya* (external), *Abhyantara* (internal), and *Bahirantara* (exter-internal). The disease in its first stage is curable ; in the second, when the joints become involved, it is cured with difficulty ; while in its third stage, when it spreads both externally and internally, the affection is pronounced as altogether incurable. One afflicted with the malady becomes lean and weak, his nose sinks down, his gastral fire becomes dim, and his bones turn dry and crooked. Mercury, catechu, *Spilanthes oleracea*, and honey in certain proportions, are recommended as a remedy. Other recipes are also given. Bhava Mishra was the first to make mention of certain medicinal drugs of countries other than India. For instance he mentions

" Badakshani Naspasi," *i.e.*, " Amrita," fruit of Badakshan.

"Khorasani Vacha," *i.e.*, *Acorus Calamus* of Khorasan ;

" Parasika Vacha," *i.e.*, *Acorus Calamus* of Persia ;

" Sulemani Kharjura," *i.e.*, date fruit of Suleman.

Bahava Mishra was an inhabitant of Benares, where he is said to have had no less than four hundred pupils.

Then followed SHARNGDHARA, son of Damodara, who wrote a treatise bearing his name. The work is divided into twenty-five chapters, and is very popular in Western India. Smaller works like "Vaidyamrita" by Bhatta Moreshvar, son of Bhatta Manek (A.C. 1627), "Vaidya Jeevana" by Lolimbraja (A.C. 1633), "Bopadeva Shataka" by Bopadeva, son of Keshava, "Vaidya Vallabha" by Hasti (A.C. 1670), "Chikitsa Sangraha" by Chakradatta, "Chikitsanjana" by Vidyapati, and others, are frequently consulted by the native practitioners.

# CHAPTER III.

## THE HINDOO THEORY OF CREATION.

THE Hindoos hesitate to give any system the name of science or *Shastra*, if it does not directly or indirectly lead to a correct knowledge of the Kosmos, and to the attainment of beatitude and deliverance from all pain and misery. The ultimate object of Medical Science is therefore stated to be to gain that knowledge which consists in discriminating the principles of the material world from the cognitive principle, the immortal soul. According to Hindoo doctrine, the whole creation is the result of the coming together of PURUSHA (spirit), and PRAKRITI (matter). The spirit is infinite, immortal, sentient and blissful. Matter is lifeless, but possesses a creative force and properties of goodness, passion and apathy. Some say that matter has no separate existence at all. It is only a manifestation of spirit, and what is known by the name of the material world is only a series of impressions of the spirit. Others maintain that matter, though helpless without spirit, is co-eternal with it; and when it comes into union with the spirit, it becomes active and procreant. It is not intended to dwell at length on the details of this system. Suffice it to say, that it recognises Man as a microcosm possessing parts corresponding to the globe, and describes him as made up of the following twenty-five principles (*tattvas*):—

1. PRAKRITI or Nature, prime cause of all things, the universal material cause (*prima matrix*).
2. BUDDHI—Intelligence, the first step in the evolution of Prakriti.
3. AHANKARA—Self-consciousness.
4. MANAS—Mind.
5. SHABDA—Sound, rudiment of ether.
6. SPARSHA—Touch, rudiment of air.
7. ROOPA—Form, rudiment of fire.
8. RASA—Taste, rudiment of water.
9. GANDHA—Smell, rudiment of earth.
10. SHROTRA—Organ of hearing.
11. TVAK—Organ of touch.
12. JIHVA—Organ of taste.
13. CHAKSHU—Organ of sight.

14. GRAHANA—Organ of smell.
15. VAK—Organ of speech.
16. PANI—Organ of prehension.
17. PAD—Organ of locomotion.
18. PAYU—Organ of excretion.
19. UPASTHA—Organ of generation.
20. AKASHA—Ether.
21. VAYU—Air.
22. TEJAS—Fire.
23. AP—Water.
24. PRITHVI—Earth.

The PURUSHA, or the Soul, is the 25th principle, which resides in the body. It is everlasting, intelligent, endless, all-pervading, blissful, immortal, calm, pure, and one without a second. These principles are arranged "in order of their development." The human organism, and for the matter of that the whole creation, thus constituted by the combination of PURUSHA and PRAKRITI, is represented in the mystical works of the Hindoos by a figure formed by drawing a horizontal line across a perpendicular one, with the ends turned round like arcs of a circle thus :—

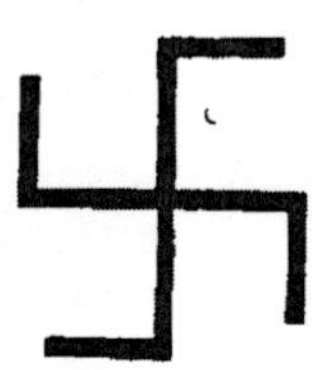

The four points of the cross represent in succession birth, life, death, and immortality, while the circle is the symbol of the eternal existence. Those who have studied the subject are trying to read the esoteric meaning of the Christian Cross in this light. It is curious that the sign of the Cross is to be found in almost all the religions of the world, ancient and modern. The Purusha is the instrumental cause of the universe, while Prakriti is its material cause. The human body is therefore believed to be the result of the joint operation of these two principles.

The creation is of two kinds, animate and inanimate. The animate creation is again sub-divided by the Hindoos into four classes, namely, *Udbhija* (sprouting), as trees, plants, etc.; *Svedaja* (produced from sweat or warmth of the earth), such as bugs, mosquitoes, etc.; *Andaja* (oviparous), as fowls and reptiles; and *Jarayu* (viviparous), as man and beasts. In the human structure, the father represents *Purusha*, and the mother *Prakriti.* When both are young and strong the offspring is healthy.

Of the several parts of the body, the hair, nails, teeth, arteries, veins, tendons and semen derive their origin from the father; while muscles, heart, blood, marrow, fat, liver, spleen and intestines owe their formation to the mother. The development of the body, its complexion, power, and condition are the products of chyle; while knowledge, perception, life, happiness, and misery are the functions that come into operation under the direct influence of the soul. The Hindoo medical works mention the possibility of a woman uniting with another woman in sexual embrace and begetting a boneless fœtus. They also believe that a woman under certain conditions may become pregnant by the influence of dreaming; and they thus explain their belief in such unnatural births as of serpents, scorpions and the like. In such a case both the woman and her production are looked upon as very sinful.

## CHAPTER IV.

### HINDOO PRACTICE DURING THE PERIOD OF NUBILITY.

A WOMAN is considered nubile during the menstrual epoch, which lasts generally from the twelfth to about the fiftieth year. During the menstrual flow she is strictly prohibited from intercourse with her husband. She is enjoined to sleep on a grass bed, to shed no tears, and to take no bath. She is not to pare her nails, and should neither run nor speak aloud. She should not apply oil or sandal to her body; and she should take care not to expose herself to inclement weather. Any disregard of these rules is regarded as being injurious to the offspring. If she cries during the monthly period, the child will contract an eye disease. The smearing of the body with oil will make the child leprous. If she sleeps during the day it will become dull and sluggish. It will become deaf if she hears a very great noise, and insane if she speaks too loudly. The period for impregnation is the first sixteen days after the appearance of the menses; of these, however, the first four days are not recommended. The best period for conception is from the fifth to the sixteenth day. Conception takes place by the union of the fecundating *Retas* (sperm) of the male with the *Rajas* (germ) of the female. It is believed that should the conception be on even days,—that is to say, on the sixth, eighth, tenth, or twelfth day,—the sex of the infant will be male; if on an odd day the sex will be female. Some are of opinion that a male child is formed when the mixture has a stronger element of semen; and that when "*Rajas*" or the germ predominates a female child is formed. If the semen virile is divided into two by the "local wind," twins result. The sex of the infant in the womb can be determined by certain signs. In the case of a male fœtus the form of the uterus is round; the right eye appears larger than the left; the right breast begins to secrete milk before the left; the right thigh becomes more plump; the countenance looks bright and cheerful; the woman desires food of a "masculine" * kind, and dreams of mangoes and water lilies. In the case of a

* In Sanskrit as well as in all the Vernacular dialects of India the gender of nouns is determined not by the distinction of sex only. Animate and inanimate objects are alike either masculine, feminine, or neuter. It is difficult to reduce the usage of the language on this point to fixed rules. Broadly speaking, things that convey the idea of largeness, strength, coarseness and firmness are masculine, and those that are smaller, weaker, finer and more delicate are said to be feminine.

female fœtus, the opposite are the signs, and the form of the uterus is ovoid. Twins are diagnosed by a median depression along the abdomen. When the sides of the woman become full, and the belly protuberant, and when the form of the uterus is hemispherical, the womb is supposed to contain an impotent.

Impotents are of five kinds, namely, Asekya, Sugandhi, Kumbhika, Irshyaka and Shanda. The last is absolutely impotent, possessing no virile power whatever. The rest are relatively so, more or less.

Barren women are divided into five classes, and are known by the names of *Kaka vandhya* (crow-barren), who bears only once in a lifetime like the crows, which are supposed to lay eggs only once; *Anapatya*, who is incapable of conceiving at all; *Garbha sravi*, who can conceive but always miscarries; *Mrita vatsa*, whose offspring do not survive their birth; and *Balakshaya*, who is sterile on account of physical weakness. All these, except the first, are capable of being cured by appropriate medical treatment.

It is particularly desirable to gratify a woman during pregnancy with everything she may conceive a wish for. For in case her wishes are not granted there is a probability of the child becoming deformed and defective. She should be kept happy and contented, should wear white clothes and ornaments, and avoid disagreeable sights and smells, and sexual and other excitements; should take easily digestible food; should neither remain hungry nor eat too much. She is enjoined not to touch a dirty, ugly, or defective woman, and is advised not to live in a lonely house or to have her bed very high.

Parturition generally takes place after nine calendar months, when the fœtus is fully developed. Sometimes the period of gestation extends to the tenth, eleventh, or even the twelfth month in exceptional cases. It is laid down that the lying-in room should be clean, and not less than eight cubits long, and four cubits broad, with ventilators in the north or the east wall. Four old and experienced midwives should be at hand to render necessary assistance. They should be trustworthy, skilled in their work, obliging, and have their nails cut close. When the time of delivery draws near, they should lubricate the genital tract with sweet oil, and one of the four should thus advise the woman in labour: "O Lady, do bear down when you are inclined to do so, but not otherwise; strain gradually, and to the utmost of your power when the child reaches the orifice, until it is expelled with the after-birth." Untimely straining makes the child deaf, dumb, hunchbacked, asthmatic, consumptive, or weak. In cases of complicated labour obstetric operations are recommended. If the child dies *in utero*, the woman feels thirsty, suffers from hard breathing, languishes, and becomes insensible. Prompt measures should then be taken to save the life of the patient. Harita recommends, among other measures, the use of a surgical instrument called *Ardha-chandra*, with which the arms of the child should be amputated and taken out and then the body. In cases of tedious labour a paste of a drug called Langali (*Gloriosa superba*) is to be applied over the hypogastrium to hasten delivery.

Those who are given to supernatural beliefs draw a double triangle interlaced, one triangle pointing upward, the other downward, with certain mystic letters written within. This figure, drawn

on a metal plate, is shown to the woman in birth-throe and placed under her bed to hasten delivery. In cases of tedious labour the medical works of the Hindoos mention certain charms and incantations which are supposed to render the delivery easy. It is curious to note that the use of charms, talismans, and incantations as remedial measures in sickness can be traced to almost all the nations on our globe.

A woman in her confinement should be most particular about her regimen. She should take no 'cooling food,' and should abstain from all bodily exertion, from sexual intercourse, and from anger. She should eat moderately and continue the necessary fomentation. Dhanvantari says that the woman's period of confinement is over after a month and a half; though she should be allowed rest for full three months. No matter requires greater attention than the quality of the mother's milk. Good milk mixes readily with water without changing colour, contains no filaments, and is white, cool, and not too thick. Milk which when mixed with water floats on the surface or sinks down, or which forms yellow spherules and is sticky and astringent in taste, is bad. A woman can improve her milk by taking green gram gruel for food and a decoction of Patol (*Trichosanthes dioica*), Nimba (*Melia azadirachta*), Asana (*Bridelia tomentosa*), Daru (*Pinus deodara*), Patha (*Cissampelos glabra*), Murvya (*Sanseviera zeylanica*), Gaduchi (*Tinospora cordifolia*), Katurohini (*Picorrhiza kurroo*) and dry ginger. If the employment of a wet-nurse be indispensable, she should be selected from the caste to which the mother belongs, should be of middle age, secreting good and sufficient milk, and having a male child. She should be of good disposition, always cheerful, exceedingly kind, obedient, contented, well-behaved, of good parentage, truthful, and willing to treat the child as her own. It is prejudicial to the child's health to be suckled by a nurse who has pendulous mammæ, or is tall, short, corpulent, thin, pregnant, feverish, fatigued, hungry, careless about food, gluttonous, sulky, mean, immoral, diseased, or suffering from pain of any kind. The medical works of the Hindoos refer to certain rites which are observed by some even now on the occasion of suckling the child for the first time. The mother has to be clean, well-clad, and to sit facing the east. She then washes her right breast and squeezes out some milk from it. The father or the priest then sprinkles a little water over the infant, reciting an incantation, the mother or the nurse keeping her hand on the right breast all the time. The incantation is to this effect:—"O Child, let the Sea of Milk * fill the Mammæ with milk for thee, and be thou strong and happy for ever. O lovely-faced lady,

* The ancient Hindoo cosmography divided the world into seven DVIPAS or continents, each surrounded by a SAGARA or sea, and the KSHEERA-SAGARA or the Sea of Milk was one of them.

let thy child live long by drinking the nectar-like milk, just as the gods are able to live for very many years by drinking the beverage of immortality." The infant is then taken by the mother on her lap, its head being kept towards the north, and nursed gently. Sushruta says that if some milk be not thrown away, as recommended above, the baby suffers from puking, cough and asthma. When the mother has no milk, and it is difficult to procure a wet-nurse, the child should be fed on cow's or goat's milk. It should always be handled gently, and never disturbed in sleep nor made to sleep against its inclination. Anointing, bathing, *Anjan* (a certain application to the eyes), and soft clothing, are always good for infants. The mother's milk may be thick, hot, acid, scant, salt, or 'soft.' The last kind is the best, and makes the child strong, healthy, and handsome. The other kinds of milk are injurious to the child and cause various diseases. A mother having scant milk may take with advantage milk mixed with black pepper and long pepper, which will promote the secretion. Similarly, powdered long pepper, dry ginger, and Haritaki (*Terminalia Chebula*) mixed with clarified butter and treacle, if taken in the form of an electuary, will assist the secernment considerably. Harita says that a preparation of dry ginger, long pepper, black pepper, the three myrobalans, Dhana (*Coriandrum sativum*), Yavani, Shatavari (*Asparagus tomentosus*), Vacha (*Acorus Calamus*), Brahmi (*Hydrocotyle asiatica*) and Bhargi, given with honey to the infant, will accelerate the power of speech and improve the voice. The memory and intelligence of the child can be greatly improved by giving it an electuary of Gaduchi (*Tinospora cordifolia*), Apamarga (*Achyranthes aspera*), Vidanga (*Embelia ribes*), Shankhapushpi (*Clitoria Ternatea*), Vacha (*Acorus Calamus*), Haritaki (*Terminalia Chebula*), dry ginger, and Shatavari, with clarified butter.

A passing reference should now be made to the several rites performed by the Hindoos from conception to delivery, and from the time of birth even until after death. These rites are twenty-five in number, but the principal ones are sixteen, called the sixteen SANSKARAS. These are: GARBHADHANA, a ceremony performed previous to conception, that is, when the husband meets his wife for the first time on her attaining maturity: PUNSAVANA, a festival held on the wife's perceiving the first signs of pregnancy; it is generally performed in the third month: ANAVALOBHANA, a rite performed to avert miscarriage: SIMANTONNAYANA, the ceremony of parting the hair of a woman on her entering the fourth, sixth, or eighth month of gestation:

JATAKARMA, rites at birth, among others putting ghee into the child's mouth with a golden spoon, before cutting the cord: NAMAKARANA, naming the little one on the eleventh, twelfth, or any other auspicious day: NISHKRAMANA, taking the infant out of the house when three months old to see the moon in the third light fortnight: SURYANI-LOKANA, the ceremony of showing the sun to the child when four months old: ANNAPRASHANA, feeding the baby with its first dish of rice in the sixth or eighth month: KARNAVEDHA, the ceremony of boring the child's ears, generally performed in odd months after birth: CHUDAKARANA, the rite of shaving the head save one lock, called "Chuda" or crest, in the first or third year, and not later than the fifth year: UPANAYANA, the investiture with the sacrificial thread, which falls from the left shoulder to the right hip, for a Brahman in the eighth and not later than the sixteenth year; for a Kshatriya in the eleventh and not later than the twenty-second year; and for a Vaishya in the twelfth and not later than the twenty-fourth: this ceremony marks the commencement of student-life: MAHANAMYA, an initiatory rite generally four days after the last, when the Gaetree* is taught and repeated:

* "The most sacred and most universally used of all Vedic prayers."

SAMAVARTANA, a ceremony on the student's completion of his studies and return home after having passed thirty-six, eighteen, or at least nine years *in statu pupillari*:* VIVAHA, marriage: the last is SVARGAROHANA, or the funeral rite. Besides the above there are other ceremonies which are performed either daily, monthly, yearly, or occasionally, the object of all of them being more or less the preservation of health both of body and mind. Any discomfort in the body is called a disease. Diseases are classed under four heads, viz., AGANTUKA (accidental), such as a fall or a cut; SHARIRA (physical), such as headache, fever, dysentery, cough, etc.; MANASA (mental), as insanity, fear, grief; SVABHAVIKA (natural), as thirst, hunger, sleep. Entire freedom from these, by the application of proper remedies, is the avowed object of medical science. The prevention of disease is considered by the Hindoos to be of greater importance than its cure. Accordingly, their medical works lay great stress on certain rules of conduct to be observed all the year round. As these precepts enable us to peep into the principles of Hygiene as understood by the ancients, it will be well now to devote some space to a short summary of them.

* This shows that the Hindoo Scriptures do not enjoin child-marriage.

# CHAPTER V.

## PRINCIPLES OF HYGIENE AS UNDERSTOOD BY THE HINDOOS.

THE importance of a knowledge of the characteristics of the country one lives in attracted the early attention of the ancients. Countries were generally arranged into three classes, namely, ANUPA, JANGALA, and MISHRA. ANUPA is a moist and marshy country intersected by numerous rivers, lakes, and mountains; and containing swans, cranes, geese, hares, pigs, buffaloes, deer and other wild animals, as well as a variety of fruit and vegetables, including paddy, sugar-cane and plantain tree. In such a country "phlegmatic" diseases and "affections of the wind" are very common. JANGALA is a dry country where water is scarce; where Shami (*Acacia farnesiana*), Kareera (*Capparis aphylla*), Arka (*Calatropis gigantea*), Peelo (*Salvadora *indica*), and jujube trees abound; where the fruits are exceedingly sweet, and beasts like donkeys, bears and spotted deer, are seen in great number. In such a country the diseases of "bile and blood" are frequent. MISHRA is a country which has all the advantages of Anupa and Jangala without their disadvantages. It is neither too moist nor too hot. Such a country is naturally the best, as it promotes health and longevity. A change recommended to a patient on the principle underlying the above classification is supposed to hasten his recovery. A man suffering from a "phlegmatic disorder" might with advantage go to a Jangala country, and in the same way one suffering from biliary complaints might profit by resorting to an Anupa country. It is the duty of the physician to preserve the health of his patient by keeping the various humours in his body in equipoise, and a knowledge of the climates is indispensable to him. After acquiring knowledge of the country, the Hindoos have got to attend to their personal duties in a prescribed manner. It is good for a healthy man to rise early in the morning, that is, about an hour before sunrise, and remember Vishnoo, the preserving power of nature. For the obtainment of longevity the names of Ashvatthama, Bali, Vyasa, Hanumana, Vibhishana, Kripa, Parshurama, and Markandeya, who are long-lived and are supposed to be still living, though ages have rolled by, are also to be recalled to memory. The first things a person should look at and touch after rising from bed are curds, ghee (clarified butter), a looking-glass, *Sarsapa* seeds, Bilva (*Ægle Marmelos*), Gorochana (a yellow pigment), and garlands of flowers; and if he desires a long life he should daily look at his face reflected in ghee. He should then answer the calls of nature with his head covered. Then he should clean his teeth with a tooth paste or powder. The substances generally used for the purpose are powdered tobacco, salt, or burnt betel-nut, or some compound preparation of drugs such as pepper, dry-ginger, long pepper and Jijbal (*Xanthoxylum rhetsa*). The most common toothbrush is a tender twig of Bavala (*Acacia arabica*); but the medical works recommend other twigs, to which wonderful properties are ascribed. A brush of Arka twig gives strength; of Vata (*Ficus indica*) gives brightness to the face; Karanja (*Pongamia glabra*) ensures victory; Pippala (*Ficus religiosa*) brings wealth; Jujube, good dinner; Mango, health; Kadamba (*Nauclea Cadamba*) sharpens memory; Champaka (*Michelia Champaca*) improves the organs of speech and hearing; Jasmine averts bad dreams; Shireesha (*Acacia Serissa*) promotes health and prosperity; Apamarga (*Achyranthes aspera*) increases patience and thoughtfulness; Pomegranate and Conessi bark improve bodily beauty; while a tooth-brush of Gunja (*Abrus precatorius*), Katala, Hintala, Brihadhara, Ketaki (*Pandanus odoratissimus*), date or cocoa tree, makes a man 'impure.' Persons suffering from certain diseases are prohibited from using the tooth-brush. After cleaning the teeth the tongue is polished by means of a scraper, which may be of gold, silver, or copper, or even of a split twig ten fingers long. Then the mouth is rinsed with cold water several times and the face washed. This process keeps the mouth free from disease. The washing of the mouth with cold water is a necessary adjuvant to remedies for aphthæ, pimples, dryness of and burning sensation in the mouth. Washing with lukewarm water removes phlegm and wind, and keeps the mouth moist. The nose is preserved from disease by dropping into it a little rape-seed oil every day. This tends to keep the mouth sweet, improves the voice, and prevents the hair turning gray. White antimony applied to the conjunctiva with a lead or zinc pencil, besides making the eyes beautiful, ensures acute vision. Black antimony of the Sindhu mountain can be used even in its unrefined state. It removes the irritation, burning, hypersecretion of mucus and painful lachrymation, renders the eyes beautiful, and enables them to stand the glare and the wind. One who is fatigued, feverish, has kept vigils or taken a meal, should abstain from applying antimony to his eyes. The nails, beard, and hair are to be kept clean and trimmed, and are to be cut every fifth day. This promotes strength, health, cleanliness, and beauty. The hair should be combed; and the looking-glass should be in constant use, as that tends to the improvement of the complexion and the prolongation of life. The hair in the nose should not be pulled out, as doing so will impair the eyesight. Regular exercise should be taken every day. It makes the body light and active, the limbs strong and well-developed, and the gastral fire increases so much that any kind of food is soon digested. Physical exercise is the surest means of getting rid of sluggishness. It is always beneficial to those taking food rich in fats. It is most wholesome during spring. Exercise after dinner or after sexual intercourse is injurious. It is not recommended at all for one suffering from asthma, consumption, and chest disease. Over-exertion is deprecated. There are various kinds of physical exercises, in-door and out-door. But some of the Hindoos set aside a portion of their daily worship for making salutations to the Sun by prostrations. This method of adoration affords them so much muscular activity that it takes to some extent the place of physical exercise.

Perfumed oil should be rubbed over the body, especially over the head, ears, and soles of the feet. Medicated oils diminish fatigue, promote strength, comfort, and sleep, and improve the colour of the skin, keep it soft and healthy, and thus contribute to the prolongation of life. The anointing of the head with oil prevents, or helps to cure, diseases of the scalp, and assists the growth of hair. Similarly, dropping oil into the ears prevents ear diseases. The eardrops, if they consist of vegetable juice, should be used before eating, and if of oil, after sunset. Oil well rubbed into the soles of the feet strengthens the legs and prevents fissuring of the skin. It also induces sleep and improves the vision. As serpents never go near an eagle, so, it is said, diseases do not approach a person who is in the habit of taking physical exercise and anointing his limbs with oil. The whole body is energised if anointed before the daily bath. But the anointment is deprecated in fresh cases of fevers, indigestion, anæmia, or vomiting. It is also to be avoided by one who has taken a cathartic. Anointment is followed by bathing. Every Hindoo is required to bathe * at

* Among Hindoos bathing is included as part of their religious duty. Manu's ordinance is :—"Early in the morning let him void fæces, bathe, decorate his body, clean his teeth, apply collyrium to his eyes, and worship the gods" (iv. 203). Yajnavalkya also recommends ablution as one of the required religious observances (iii. 314). As a rule, bathing is a pre-requisite to the morning meal, though not a few of the higher classes perform ablution before taking their evening meal also, as well as after touching any unclean thing. Hindoos boast that they are the most cleanly nation in the world, and this statement is borne out by a remark of Sir William Hunter, who says: "It is needless to say that the Indian Hindoos . . . . . stand out as examples of bodily cleanliness among Asiatic races, and, we may add, among the races of the world. The ablutions of the Hindoo have passed into a proverb. His religion demands them, and the custom of ages has made them a prime necessity of his daily life."

least once every day. Bathing after a meal is injurious. A cold bath is a preventive of blood-diseases, while a hot one has an alterative effect. The daily use of an emblic myrobalan bath preserves the black colour of the hair and ensures life for a hundred years. Too hot a bath is injurious to the eyes. To bring the water for ablutions to the required temperature, it is directed that hot water should be added to cold, but that cold water should never be added to hot water. An old physician named Harishchandra says: "O men, a warm bath, fresh milk, a young damsel, and moderate use of fatty articles of food, are conducive to your health." Persons suffering from ague, cold, diarrhœa, dyspepsia, ear affections, or eye diseases, should abstain from bathing. When the bath is over, the body is to be carefully rubbed dry with a towel and properly dressed.

It may be noted by the way that the Hindoos, as a rule, never bathe in a nude state either at home or in public. There is a religious interdiction against exposure of the person (Manu, vi. 45). In cold weather saffron, sandal, and black aloes are applied to the body; in summer, a paste of sandal, camphor, and *Andropogon muricatus* is recommended, and in the rainy season the body may, with advantage, be smeared with a preparation of sandal, saffron and musk. Good men are advised never to put on dirty clothes, as they cause irritation and other diseases of the skin, and make one look disreputable. One should wear flowers and ornaments according to one's taste and means. Fragrant flowers and leaves beautify the body, excite amorous passion, and drive away evil spirits. Gold is holy, auspicious, and a giver of contentment. Precious stones possess the efficacy of averting evil eyes and evil influences of the planets, as well as bad dreams and wicked intentions. Charaka adds that, after the purification of the body and before meals, it is proper to devote some time to the worship of the Almighty.

A man should take his meals twice a day—in the morning between nine and twelve o'clock, and in the evening between seven and ten. The meals should not be taken in a public place, as it is said that eating, cohabiting and answering the calls of nature should always be done in private. A dinner-service of gold is the best from a medicinal point of view, and it is supposed to be the best tonic for the eye. Eating out of silver is equally efficacious in promoting hepatic functions. A service of zinc improves the intelligence and appetite. Food served in brass utensils promotes wind and heat, but cures phlegmatic disorders and expels worms. The use of steel or glass vessels cures chlorosis, jaundice and intumescence. A stone or clay service brings on poverty. Wooden plates are good appetisers, but help the secretion of phlegmatic humour. The use of certain leaves as plates acts as an antidote against poison. When at dinner, a water jug with a cup should be placed on the right hand. A copper vessel is the best for the purpose. The next best is an earthen pot. Vessels made of crystal and lapis lazuli are also pure and cooling. It is good to take a little rock-salt and fresh ginger before entering the dining-room, as this is supposed to whet the appetite and clear the throat. Charaka says that one should not sit to dinner facing the north. Manu's dictum on this point is somewhat different. He says that one desiring longevity should face the east while having his meal; one desirous of fame must face to the south; of wealth, towards the west; and one desiring true knowledge should sit looking towards the north (ii. 52). Should one happen to pass flatus during a meal he is to leave off eating and not to take any food during the day. The name of "Hanumana, son of Anjanee" is mentioned to avert the influence of evil eyes, and also the name of the Supreme Being, who "is the Fire residing in the bodies of living creatures, where, joined with the two spirits which are called Prana and Apana, He digests the food which they eat, which is of four kinds" (Bhagvat Gita, xv. 14). The four kinds of food above referred to are (*a*) those to be masticated with the teeth, as bread; (*b*) those licked with the tongue, as chutney; (*c*) those sucked in with the lips, as mango; and (*d*) those simply imbibed, as liquids. The various dishes are served one after another in a prescribed order and are put in the places assigned to them. The food placed before one is to be treated with divine respect, such treatment being conducive to health and strength (Manu, ii. 55). Pomegranates, sugar-cane, and similar things, should be eaten first and never at the end of dinner. It is good to take hard and butyraceous substances in the beginning, soft viands in the middle, and the liquids towards the end of the meal. Similarly, sweets are to be taken first; salt and acid things next; and pungent, bitter, and astringent things at the end. The dinner should be finished with a draught of milk, or Takra (whey) mixed with water. One should not hurry over his meals. Gormandism is to be avoided. Half the cavity of the stomach is to be filled with food, a quarter with water, and the remaining part is to be left empty. Water may be taken now and then during the meal; if taken in the beginning it retards digestion, and has a tendency to make one lean; if taken at the end it produces obesity (Vagbhata). A thirsty man should not eat before quenching his thirst, and a hungry one should not drink before taking some food. Any disregard of the first rule causes tumour, and of the second dropsy. Sushruta draws particular attention to the advantages of dining at fixed hours, and recommends that the food once taken off the stove should never be heated over the fire again. One must study the nature of the food before eating it, for the food one eats has much to do with the development of the mind, and it is the mind that makes a man either good, bad, stupid, or wicked. "There are three species of food dear unto all men. The distinctions are based on the inherent quality or GUNA of the food. The food that is dear unto those of the SATVA GUNA (quality of goodness) is such as increases their life, their power, and their strength, and keeps them happy, contented, and free from sickness. It is pleasing to the palate, nourishing, substantial, and congenial to the body. The food that is coveted by those of the RAJO GUNA (quality of passion) is either very bitter, sour, salt, hot, pungent, astringent, or very heating, and giveth nothing but pain and misery. And the delight of those in whom the TAMO GUNA (quality of darkness) prevaileth, is such as was dressed the day before and is out of season; has lost its flavour and has gone putrid; the leavings of others and all things that are impure" (Bhagvat Gita, vii. 8–10). Those who desire to have the quality of goodness should take the food used by the "Satvikas" and not others. When the meal is over, the mouth is scrupulously cleaned, both inside and out, by water, so also the hands. Salt may be used to remove the greasiness. Any particles of food sticking between the teeth should be picked out. The eyes should be gently stroked with the wet hands, as this has the effect of improving the vision. Then a prayer is offered to Agastya, Agni, and Vadavanala (the Submarine Fire which is supposed to devour the waters of the ocean) to the following effect: "O help me to digest the food I have eaten; let me have the happiness resulting from well-digested food; and relieve me from all diseases." Mangala, Surya, and Ashvinikumaras are also piously remembered, as the mention of their names is said to possess the power of helping the digestive organs. After dinner, aloe-smoking or the chewing of Pan (betel-leaf) with certain aromatics and spices is advisable, for it has the property of expelling the phlegm which increases after dinner. The Pan is astringent, exhilarant, aromatic, stimulant, carminative, aphrodisiac, 'light,' and heating. It is a good phlegmagogue, generates semen and blood in the body, and lessens wind and fatigue. The various ingredients mixed in certain proportions with the betel-leaves are catechu, lime, betel-nut, cardamum, clove, nutmeg and some other spices. In the masticatory to be taken in the morning the quantity of betel-nut may be a little more than at other times; at noon catechu may be a little in excess; and at night the proportion of lime may be a trifle more. It removes all fetor from the breath, imparts fragrance to it, and improves the voice. The Pan is not beneficial to those who are suffering from tooth and eye diseases, who have taken an opening medicine, or who are in an intoxicated state. Walking a hundred paces after dinner promotes life, while remaining sitting brings on idleness. Running after taking one's meal is tantamount to running after death. It must therefore be avoided. After a brief lounge, the best thing to do is to lie down on the left side for a while, as this position favours digestion. At this time the Hindoos generally undergo the process of shampooing—which is but another name for massage,—flexion, extension, rotation, pronation, supination, adduction, abduction and circumduction of the various parts of the body, as well as racking the joints and employing gentle blows and friction, forming part of the manipulation. This purifies the flesh, blood, and skin, exhilarates the mind, brings on sleep, cures diseased phlegm, wind and fat, diminishes fatigue, and increases internal heat. The practice is peculiar to Hindoos, and is referred to in their ancient works. Shampooing in one form or another has been practised from immemorial ages by the Chinese, the Greeks, and the Romans, who, according to some Western authorities, seem to have obtained its knowledge from the Hindoos. It is of various kinds, and the barber caste in India is supposed to be expert in the art. It is a great curative agent in the treatment of complaints connected with the nervous system, and always produces the most agreeable sensation. The advantages of shampooing have begun to be appreciated by the Western Medical Science, which no longer hesitates to recognise massage as a therapeutic agent. This fact is viewed with satisfaction by the Hindoos, who fondly hope to see in the several scientific discoveries of the West the revival and salvation of their own medical lore, which, instead of being treated as empirical, will be recognised as the collective wisdom of those who have had due regard to science and theory. In India, people—especially the males—are in the habit of being shampooed more as a matter of luxury than anything else. Female patients are operated upon by female experts only. Excessive indulgence in massage, as in everything else, is deprecated.

Sleeping in the day-time, except in summer, is discouraged. It is allowed to those who are given to walking and riding long distances, and who are in the habit of undergoing much physical exertion; also to children and the sick, as well as to those who can "control their sleep," that is, those who can dispense with sleep at night if they sleep during the day. Immediately after dinner it is highly injurious either to bask in the sun, sit by the fireside, swim, ride, run, fight, sing, take physical exercise, or study. A sensible man should never have sexual intercourse in the day-time, as that shortens life. After taking a little rest, one may engage himself in his daily avocation. An evening constitutional is particularly recommended, as it makes the senses active, excites the action of the stomach and the skin, and improves the intelligence. When going out, the head should always be protected with a light turban, and the feet with shoes. It is not safe to put on shoes, clothes, and garlands used by others.* An umbrella may be used in the hot and rainy seasons. One should never be without a walking-stick, as it protects him against beasts, prevents fatigue, and "adds dignity to the individual." He should not look at the reflection of himself in water, nor should he enter the water stark naked. One is to be always industrious, and should never neglect the calls of nature. An old

* This advice shows that the Hindoos were not blind to the risk of contagion.

man, a pandit, a doctor, a king and a guest should always be respected. The organs of sense should be neither overtaxed nor allowed to remain idle. It is harmful to see the rising or the setting sun, to carry any burden on the head, or to sleep on a torn bed or under a tree.

Such a line of conduct is conducive to long life, health, and fame. Having taught how to behave during the day-time, the ancient Hindoo writers on medicine have laid down rules of life to be observed during the night. They direct that dinner, cohabitation, sleep, study and walking in the street, are not to be indulged in at sunset. The chances of enjoying the moonlight should not be missed, as it is cool and soothing, and increases the sexual appetite and powers. Supper should always be light. Curds are to be avoided at night, and must never be used without the addition of some salt. Sexual intercourse should be in moderation only. With the Hindoos the object of the marital relation is not so much the gratification of the animal passion as the fulfilment of an obligation. It is enjoined on them to beget a progeny—a *putra* (son), or a *putri* (daughter). The word "putra" is derived from "pu," hell, and "tra," to liberate, and means one who can liberate the Manes from hell. For the common belief is, that as long as one does not have offspring, especially male offspring, his Manes are doomed to perdition. One dying without a son is offered no salvation. Marriage among the Hindoos is therefore a religious sacrament and not a social contract. To beget a son is, with them, to liquidate the debt they owe to their ancestors. No one would like to be called childless, as that is equivalent to a frustration of the real object of matrimony. Sometimes, owing to disparity of age between the husband and wife, or owing to defects in the generative organs of the one or the other, or both, a successful insemination is not possible. With a view to remove these disabilities the sage Vatsayana, an author who wrote about the beginning of the Christian era his book called "Kamasutras," or "Aphorisms of Love," prescribes some remedies. He alludes in his writings to the works of seven earlier authors on the same subject. His disciple Koka has earned a wider popularity. He describes the various causes that prevent conception, and recommends remedial measures. Among other remedies he lays particular stress on Posture, which, according to him, has great influence on the female pelvic organs; and he indicates certain positions as facilitating impregnation and curing internal disorders.* He describes as many as eighty-four positions or ASANAS, which may be resorted to under varying conditions, and adds that these postures not only heighten the pleasure of the moment, but act as a means of ensuring fecundation. Owing to the extreme delicacy of the subject treated of by this writer, his work *De Rebus Veneris*, though translated into several languages of Asia, appears always to have been held in doubtful repute. But, apart from a layman's point of view, it deserves to be appreciated by the medical profession, which has only recently recognised Postural Treatment as a new and useful therapeutic method in Gynæcology. The subject is still in its experimental stage; but when the time comes for a universal recognition of Posture as a curative agent, the Medical Science of India will be able to claim the credit of having been the first to propound the theory. The medical works of the Hindoos have from the

* Cf. *Lucretius*, Book IV.

earliest period recognised the influence of Posture in parturition, and described in detail the positions which women in labour should adopt. A sage named Patanjala, the founder of the Yoga philosophy, who flourished about B.C. 200, in his work called "Yoga Sutras," prescribes various ASANAS or Postures for preventing and curing diseases to which ascetics and others practising abstract meditation and seeking seclusion from the world may be subject during the performance of physical austerities. It will be clear from these facts that the Hindoos were not ignorant of the wholesome effect of Posture.

Sexual intercourse is prohibited for the first four days after the appearance of the menstrual flow, as well as on the 8th, 14th, and 15th days of both the fortnights—light and dark; on the anniversary days of dead parents, nights previous to the anniversaries; on VYATIPATA (the seventeenth of the astrological Yogas), VAIDHRATA (the twenty-seventh astrological Yoga), SANKRANTI (the passage of the sun or planetary bodies from one sign of the zodiac to another); in the daytime, at midnight, and during an eclipse. One authority advises men to avoid flesh, honey, oil, and woman's company on Sundays, if freedom from disease be desired. It is also said that "putrid flesh, old women, the autumnal sun, half-curdled milk, and morning cohabitation and sleep are fatal." Again, Sushruta is of opinion that the carnal desire may be gratified at the interval of a fortnight in summer, and at the interval of not less than three days in other seasons. Those who have eaten a heavy meal, are hungry, thirsty, impatient, boyish, old, with aching limbs and pressed with the calls of nature, should abstain from the indulgence. It is not proper to peep into the privacy of a bed-chamber. But the ancient writers on Medicine and Religion have not omitted to prescribe rules of conduct to be observed even there. Hiranyakeshi advises the housewife to light the lamp, and keep the bed in good order. She should make a bow to her husband, and approach the bed after removing her bodice. The retention of the bodice is supposed to bring on widowhood. She is to exclude black apparel, which has the effect of making the progeny wicked and degenerate. She is to put on clean clothes, deck her body—the nose especially—with jewels, to apply cohol to the borders of her eye-lids, and red oxide of lead (*Sindura*) to her forehead, and chew the Pan mixed with the usual spices. Both the husband and wife should be in most cheerful spirits. There should be no sign of pain or sorrow on the face of either. The wife is then to wash the feet of her lord, rub fragrant powders over his body, and burn incense before him. She places before him milk boiled with sugar, nutmeg, saffron, almond and musk to drink, and herself drinks what is left. She then offers him betel-nut and various spices wrapped in a betel-leaf, and then rests her head on his feet, takes him for her God, and calls to mind the names of worthy men that have flourished in the family, or of any celebrated sage or warrior or holy person. The husband also remembers his Creator, and prays to be blessed with a good child.* He then indulges in coition when the breath is flowing through his right nostril. For it is said that "dinner, evacuation of the bowels, cohabitation, sleep, interview with kings, fighting and taking medi-

* All this shows that the Hindoos believe in the influence on the offspring of the mental impressions of the parents at the time of its conception, and recognised, ages ago, Genætology, or the science of begetting healthy and beautiful children, which is just beginning to receive attention in other countries.

cine, should be done when the breath is passing through the right nostril." After intercourse, it is beneficial to bathe, or at any rate to wash the hands, feet, and other parts, drink soup or milk, eat articles of food mixed with treacle, open the windows, and go to sleep. The lamp should be extinguished by the wife, who is then to occupy a separate bed. One should not sleep with the head towards the north. Sleeping with one's head towards the south is supposed to prolong life. One passes a dreamy night by keeping his head towards the west, and gets wealth by keeping it towards the east. A sound and quiet sleep is secured by muttering the names of the "Five Happy Sleepers," namely, Agasti, Madhava, Muchakanda, Kapila, and Astika. Nothing is so beneficial as to go to sleep regularly and rise early. If a man cultivates the habit of drinking eight *anjalis* (a measure formed by putting the hands together and hollowing the palms) of water every morning at sunrise, he will be free from the effects of old age and such diseases as hæmorrhoids, inflammations, headache, shooting pain, and bilious affections, and will live for a hundred years. If one is accustomed to drink a small quantity of water through the nose instead of through the mouth, his eyesight will improve, and his hair will not turn gray.

The above precepts may be modified a little with the change of seasons. India has the advantage of enjoying six seasons, each with a regular duration of two months. They are :—

SHISHIRA, the dry season (roughly January and February).

VASANTA, Spring (March and April).

GREESHMA, the hot season (May and June).

VARSHA, the rainy season (July and August).

SHARAD, the sultry season (September and October).

HEMANTA, the frosty season (November and December).

During the first three seasons the sun remains to the north of the equator. The effect of the sun on vegetation at this time is not of the best. He is supposed to absorb the juices of medicinal herbs and impart to them heating properties. In the remaining three seasons the effect of the sun's rays on the herbaceous plants is very beneficial, and the vegetables produced in this part of the year possess cooling properties.

In SHISHIRA, when the climate is cold and dry, the morning meal should never be neglected, and pungent, acrid and salt things should be particularly used. The body should be smeared with oil; before bath, physical exercise is particularly recommended in this season. Wheat, jaggery, rice, Masha (*Phaseolus radiatus*), meat, new grain, sesamum and massage are highly agreeable. Saffron and musk may be applied to the body. The clothing should be warm.

VASANTA promotes phlegmatic diseases, so emetics may be taken with advantage in this season. Bodily exercise is also beneficial. Dry, pungent, light and heating substances are to be selected for food, and sleep in the day-time should be avoided. The season is generally unhealthy, and in it the physicians drive a roaring trade. Tepid baths are advantageous. Wheat and rice used for food should be a year old. The hottest part of the day may be spent with profit in a garden abounding in flowers and verdure, which obstruct the direct rays of the sun.

In GREESHMA the sun absorbs the phlegm secreted in the body. It is therefore advisable to eat such articles as may make up for the loss of the phlegm. Sweet, oily, cooling, light and liquid things are recommended. Sugar, curds, soup and milk may be freely used. A noon-day nap is a good prophylactic in this season. Moon-light is healthful. Pungent, salt and acid articles should be shunned. Indulgence in athletic exercise as well as shampooing is deprecated.

VARSHA gives rise to wind complaints. As palliatives, sweet, sour, and saline substances should be used for food. Sitting near the fireside is profitable, and shampooing is good. Curds should not be taken without being mixed with black pepper. Wheat, rice and Masha are good to eat. Well-water or rain-water may be used. Humidity and exposure to the east wind, or to the sun's rays, should be guarded against; so also siesta, fatigue and swimming. Sleeping on the ground floor is not advisable in this season.

SHARAD gives rise to bile distempers. Clarified butter, milk, white sugar-cane, game, wheat, barley, kidney-bean and rice may be selected for food. Sweet, astringent and bitter things should be preferred. Rain-water, and water which is exposed to the sun's rays* in the day-time and moon's rays at night, should be used for drinking

* Evidently the Hindoos fully appreciated the purifying influence of the solar rays.

purposes, and the water as a rule should be fetched in the morning. The use of camphor, sandal-wood and light clothes is recommended. Flowers, moonlight, playing in the water and light and cooling articles of food are salutary. On the other hand, curds, exercise, sour, pungent, hot and acid things, and exposure to the sun, are injurious. This is the most unhealthy season in India, and is aptly described by a common Sanskrit hemistich, "*Vaidyasya sharadi mata, pita cha kusumakaras,*" which means "the autumn is the mother, and the spring the father, of the physician." For the Vaidyas are never so busy as in the two unhealthy seasons, which provide them with the means of livelihood. "May you live a hundred *Sharads*" is a common form of benediction among the Hindoos. Purgatives to evacuate the bile, and bloodletting in strong persons, are conducive to health in this season.

In HEMANTA the rules of conduct to be observed are similar to those prescribed for SHISHIRA.

These practical precepts have received the seal of sanction and approval from the Hindoo religion, which has made them binding on the people, who still cling to them, though foreign invasions and intestine dissensions have materially affected their other social habits and their political influence. History makes mention of no other nation that has survived so many counteracting forces. If Megasthenes, who wrote about India in B.C. 800, or Hiouen Thseang, the Chinese pilgrim, who graphically describes his experiences of India in the 7th century A.C., were to rise from their graves and revisit the country, they would scarcely have occasion to alter their first impressions about the manners, customs, and the daily practices of the Hindoos. This proves pretty clearly that the various observances and hygienic directions prescribed for the guidance of the Hindoos are based on too solid a foundation to be wholly destroyed or radically affected by the ravages of time. By their daily and seasonal practices the Hindoos are directly and indirectly defending themselves against the approach of diseases. But diseases often do come in spite of preventive measures. Their medical works, therefore, prescribe remedies for curing them. Their theory of the nature of diseases is somewhat different from that recognised by modern science. But it has the merit of being original. And as it has been in vogue for centuries, it will be well to describe it briefly in a separate chapter.

# CHAPTER VI.

## THEORY OF INDIAN MEDICINE.

INDIAN Medical Science attributes all morbid phenomena to the disordered condition of the three principal humours in the body, called DOSHAS, viz., wind, bile and phlegm. These fluids pervade the whole microcosm of man. So long as these are in their normal condition the body remains healthy. If they be deranged they subject it to all sorts of disorders. The three humours fill the whole body which they support; yet the principal seat of wind (VATA) is between the feet and the umbilicus; of bile (PITTA), between the umbilicus and the heart; and of phlegm (KAFA), between the heart and the vertex. Wind predominates in old age, bile in middle age, and phlegm in childhood. Evening is the time for the predominance of wind, and noon and morning for the prevalence of bile and phlegm respectively. Similarly, the influence of wind is great after the food in the stomach is digested; when the action of the stomach is half done, or when the food is in a semi-digested state, bile gets the ascendency, and phlegm holds sway in the commencement of the process of digestion. When wind predominates, digestion becomes irregular; when bile is abundant, it is accelerated; under the controlling influence of phlegm, digestion becomes weak. For perfect digestion the three humours must be in their proper proportion. If wind is predominant the bowels become costive; when bile is in excess they become loose; when phlegm predominates, the bowels remain in their normal condition. A proper equilibrium of the three alone keeps the body healthy. Sometimes defect in the humours is congenital. In that case bilious diathesis is considered better than windy, and phlegmatic better than either, though, on the whole, any disorder in the humoral functions is undesirable. The cardinal humours, VATA, PITTA and KAFA, are expressed in English by wind, bile, and phlegm respectively; but they convey more meaning than their English equivalents are capable of expressing, as will appear from a short description of each.

Every movement of the body depends, according

to the Hindoo theory, on VATA, which alone possesses motive power. It is susceptible of taking on qualities by contact, but it is naturally dry, light, cool, sharp, fine and motive. It is of five kinds, distinguished from one another according to the functions they perform in the organism. Their names are UDANA, PRANA, SAMANA, APANA and VYANA.

UDANA is situated in the neck, above the sternum. It is by this wind that one can speak, sing, and utter sounds. When it becomes defective, it produces diseases in the parts above the clavicles.

PRANA is situated in the chest and passes through the mouth and nose, and is the means of respiration and performing deglutition. When it is deranged it produces hiccough, asthma, etc.

SAMANA is in the stomach, in the neighbourhood of the gastric fire. It converts the food introduced into the digestive canal into a nourishing juice, and separates the juice from the refuse which is to be rejected from the body. When vitiated it causes dyspepsia, diarrhœa and colic.

APANA is located in the hypogastrium. Its function is to expel fæces, urine, semen, menstrual fluid and the fœtus. When vitiated it causes constipation, diseases of the rectum, urethra, bladder and seminal disorders.

VYANA pervades the whole body, and energises it by conveying the fluids over the different parts. It produces the flow of sweat and blood, and the various movements of the body are all dependent on it. Any derangement of it gives rise to all sorts of bodily complaints. "If all the five kinds of wind are diseased the body perishes." Some writers recognise five more vital airs, and call them NAGA, KOORMA, KRIKALA, DEVADATTA, and DHANANJAYA, their respective functions being eructation, nictation, sternutation, yawning and inflation of a corpse.

PITTA is naturally hot, liquid, yellow, bitter, but acid when vitiated, light and oily. It produces animal heat, and is of five kinds.

PACHAKA—its situation is between the stomach and the small intestines (*Pakvashaya*), which are the seat of the fire of digestion. It assists digestion and imparts heat to the whole body and separates the nourishing juice (*Rasa*) and dejecta. Native writers do not seem to be unanimous in their opinion about the nature of the "fire of digestion." In the opinion of some, this bile and the bodily fire are identical; others think differently. The author of the "Rasa-pradipa" describes this fire as an exceedingly minute heating substance situated in the middle of the navel. It communicates heat to the bile and digests the food received in the stomach. In the largest animal it is no larger than a barley corn; in smaller animals it is as small as a sesamum seed, while in worms and insects it is as minute as the point of a hair.

RANJAKA remains in the liver and the spleen, and imparts redness to the essential juice, which then becomes blood.

SADHAKA is in the heart. It sharpens the memory, the intelligence, and the understanding.

ALOCHAKA has its seat in the eyes, and supports the power of vision.

BHRAJAKA is situated in the skin, to which it gives brightness and a healthy colour. It absorbs applications made to the skin, and improves the complexion.

KAFA is white, heavy, oleaginous, viscid, cooling and sweet, but becomes salt when defective. It is of the following five sorts, according to the locality in which it is situated:

KLEDANA is in the stomach. It moistens the

food, and strengthens the different organs of the body.

AVALAMBANA is situated in the heart, the shoulder-joints, and the *trik* (sterno-clavicular joints).

RASANA is in the throat and the tongue, which it keeps moist, and by means of it we discriminate the tastes of different kinds of food.

SNEHANA is in the head, and refreshes the organs of sense by keeping them moist.

SHLESHANA is situated in the joints, which it lubricates and keeps ready to perform their actions.

It is easy to find out from certain signs as to which of the humours is in excess in a particular individual. For instance, a person constitutionally subject to excessive wind is generally dark, lean, has dry and scanty hair, is susceptible to cold, garrulous, jealous, impatient, in the habit of keeping awake, walks fast, is not very fond of women, and has few children. He often dreams of flying or climbing. Vagbhata says that the dog, the hare, the camel, the vulture, the rat, the cow and the owl, are by nature subject to wind humour.

A person with bilious temperament is fair, lean, red-eyed, prematurely gray, timid, intelligent, irritable, enterprising, proud, loving self-praise, kind-hearted; a huge eater, often feeling thirsty and hungry, fond of scents and flowers, sweet, bitter, astringent and cold food, and spirits distilled from molasses; has good memory, and dreams of fire and lightning. The tiger, the monkey, the cat, the wolf and the spider, are said to be bilious by nature.

A phlegmatic person has a fair complexion, long and black hair, broad chest; likes bitter, astringent and hot diet; is strong and forbearing, true to his word, courteous, pious, and intelligent, but slow in work; is fond of vocal and instrumental music as well as lecherous, takes delight in physical exercise and is constant in love. He often dreams of rivers and ponds. The eagle, the swan, the lion, the horse and the ox, are said to have phlegmatic constitutions.

Wind is engendered by fasting, watching, jumping, severe exercise, and excessive indulgence in sexual intercourse; bile by very hot, dry and bitter food, and intoxicating drinks, as well as by anger and excess in venery; and phlegm by want of sleep, sleeping in the day-time and eating without appetite.

Besides the three humours described above,

seven more essential parts or supporters of the body are enumerated, and are called Dhatus, or the constituent parts. They are Rasa (lymph-chyle), Rakta (blood), Mansa (flesh), Medas (fat), Asthi (bone), Majja (marrow) and Shukrà (semen). Their respective functions are to cause pleasure by circulation, to energise, to plaster, to lubricate, to support, to fill the cavity of the bones, and to propagate.

As has been stated above, the RASA, which permeates the whole body by circulating through the Dhamanies, is a nutritive fluid extracted by intestinal absorption from the food which has been subjected to the action of the digestive organs. It is purely white, sweet, and cooling, and keeps a man in good spirits. When the fluid in its circulating course enters the spleen and the liver, its white colour turns to red, and it is then known by the name of "blood." Blood is formed into flesh, flesh into fat, fat into bone, bone into marrow, and marrow into semen. Rasa, when defective, becomes acrid or acid, and engenders diseases, sometimes poisoning the whole system.

RAKTA (blood), which is heavier than Rasa, also circulates in the vessels assigned to it. This theory of the motion of the blood through the different vessels of the body is worthy of attention, for it sets up the ancient Hindoos as claimants for the honours given to William Harvey for discovering the circulation of the blood in 1628. Harvey, no doubt, was the greatest experimenter of his age, and deserves the highest credit for leaving a "glorious legacy" to modern Physiology by scientifically explaining the theory of the circulation of the blood. But it is possible for him to have received his inspiration from earlier writers, who have taught something similar, if not with so much precision. If the ancient Hindoo writers on Medicine do not make mention of the circulation of the blood as frequently and explicitly as they do of the circulation of the Rasa, it is because there is, according to them, little difference between the two except in their colour and specific gravity. Both are fluid, but Rasa is a finer liquid, which supports the body and is the very essence of existence. Blood, minus its colouring ingredients, is Rasa. It may be called Chyle, or rather Lymph-chyle, though the Hindoo writers give it a wider significance than the English word is capable of bearing. The function of circulation is common to both the fluids. For Rasa, it is distinctly stated that from the heart it is propelled by the Vyana Vayu to circulate through the arteries and veins; and that it nourishes the body, as water conveyed through the canal irrigates the field. This to some extent answers the description of the circulation theory. The circulation of the blood is also mentioned by several early writers, who each and all ascribe the property of *Chala* (motion) to the blood. Harita, in his work called the *Harita Samhita,* which some believe to be older than *Sushruta,* refers to the circulation of the blood in describing a disease called "Panduroga" (Anæmia). He says that this disease is sometimes caused by swallowing clay, which some persons are in the habit of doing. "The clay thus eaten blocks the lumen of the several veins and stops the "*circulation of the blood.*" The author of *Bhavaprakasha,* who is a century older than Harvey, quotes the following couplet bearing on the circulation of the blood :—

*Dhatoonam pooranam samyak*
*Sparshajnanam asamshayam,*
*Svashirasu charad raktam*
*Kuryach chanyan gunan api.*

"Blood, by circulating through its vessels, fills the Dhatus well, causes perception, and performs other functions (of nourishing and strengthening)."

Again :

*Yada tu kupitam raktam*
*Sevate svavahas shiras,*
*Tadasya vividha roga*
*Jayante raktasambhavas.*

"When defective blood circulates through its vessels it causes many blood diseases."

Similar passages can be transcribed from even earlier writers. But the above quotations are enough to satisfy a casual reader that the circulation of the blood was not unknown to the early Aryans.

MANSA (flesh) is blood digested by heat and thickened by wind. This also produces PESHI (muscle). There are five hundred muscles in the body of a male. In females there are four hundred and ninety-seven.

MEDAS (fat) is produced by the digestion of the flesh by the internal fire. Its principal seat is in the abdomen.

ASTHI (bone) is fat digested by the internal fire and thickened by the wind. According to Sushruta, there are 300 bones in the body. Charaka includes the cartilages of the ears, eye-lids, nose, and the larynx, and makes the number 306. Of these 120 bones are in the limbs—the arms and legs—118 in the trunk, and 63 in the head and the neck.

MAJJA (marrow) is situated within the bones, and gives a shining appearance to the body.

SHUKRA (semen) is formed in males by the essential parts of marrow mixed with blood. It is the support of the body and the root of pregnancy. In the female the Rasa is converted once a month into menses, the analogue of semen in the male. When conception takes place, the menstrual fluid is diverted to the mammary glands and forms milk. Urine, fæces, sweat, cerumen, free extremity of the nails, hair, expectorations, tears, chassie and nasal mucus, are considered impurities of the body.

There are six ASHAYAS or hollow viscera for holding phlegm, undigested food (*ama*), bile, wind, fæces and urine. A female has three more for holding the fœtus and milk. Seven smaller viscera, for holding some of the essential parts, are called KALAS or receptacles. The human body contains 210 joints or SANDHIS, of which 68 are movable and the rest are immovable. There are 68 joints in the upper and lower extremities—all movable—59 in the trunk, and 80 in the head and neck. The joints are bound together by 900 SNAYUS (ligaments), thus distributed : 600 in the upper and lower extremities, 230 in the trunk, and 70 in the head.

There are altogether 700 vessels in the body, with sixteen larger ones called KANDARAS, and twenty-four called DHAMANIS. Wind, bile, blood and phlegm have each a number of vessels assigned to them. The human body also comprises sixteen JALAS or plexuses, six KOORCHAS (larger glands ?), four RAJJUS or chordlike structures, seven SEEVANIS or sutures which should never be bored, fourteen bone-groups, fourteen SIMANTAS or supporters of the groups and seven layers of skin. The names of the seven layers are:—

AVABHASINI, containing the vessels. Its thickness is one eighteenth of a barley corn. It is so called because it "shines" by the bile called Bhrajaka.

LOHITA (blood-red) is the sixteenth part of barley in thickness. It is in this layer that pimples originate.

SHVETA (white) is of a white colour, and is one-twelfth part of a barley grain in thickness. It is the seat of cutaneous eruptions.

TAMRA (copper-coloured) is a membrane an eighth part of a grain of barley in thickness.

VEDINI (sensible) is the thickness of the fifth part of a grain of barley. Erysipelas begins in this layer.

LOHINI has the thickness of a barley corn and is the seat of tumour.

MANSADHARA (flesh-holding) is the thickness of two barley corns. It retains the muscles in their places and is the seat of boils.

These layers are distinguishable only in the region of the belly and a few other parts.

A man's body has nine orifices—two nostrils, two ears, two eyes, the mouth, the anus and the meatus urinarius. The female has three more, namely, the openings of the lactiferous ducts and the orifice of the vagina.

Hindoo anatomy recognises certain MARMASTHANAS (vitals?) in the body which are most essential to life and to sound health. They are to be carefully preserved against all injury, and are arranged in five groups according to their regions and the consequences they produce when wounded: (*a*) parts which if wounded cause immediate death: there are nineteen such; (*b*) those which if injured cause a lingering death: there are thirty-three of this kind; (*c*) such as impair the limbs if wounded: there are forty-four of these; (*d*) parts which when slightly wounded produce intense pain: the number of such parts being eight; and (*e*) vital parts which produce fatal results if foreign bodies located therein be extracted: of these there are three in the body. All these parts are described at length, and surgeons are particularly warned to avoid operations on these.

The science of Aryan Medicine is, as we have seen, based on the three morbific diatheses. These dispositions are born with man—nay, it is asserted that there is no substance in the universe which does not owe its formation to the humours in more or less proportion. The humoral pathology of the ancient Aryans has been in existence for ages. Diagnosis made on the principle of this theory, and medicines administered in conformity with its teachings, have, say the Hindoos, worked pretty successfully in India. This theory seems to have been borrowed from the Hindoos by Hippocrates (460 B.C.), the Father of Greek Medicine, and to have retained its hold on the medical schools of Europe for more than 2000 years. To discard the theory as thoughtless and barbarous is, urge its advocates, unjustifiable. The epithets are strongly resented by the Aryan physicians, who complain that their science has not been properly studied and examined by modern investigators, who have condemned it on insufficient data. They are, however, taking comfort in the hope that modern medical science, in the course of its onward march, or on reaching its goal of progress, may possibly land its votaries on the very theory which they have at present rejected.

# CHAPTER VII.

## INDIAN MATERIA MEDICA.

THE Materia Medica of the Hindoos is a marvel to the modern investigator. In it are fully described the properties of drugs belonging to the animal, vegetable, and mineral kingdoms, and of the articles of food essential to the maintenance of health and strength. The theory which forms the basis of the investigation is, that every substance, whether animal, vegetable, or mineral, possesses five properties, namely, RASA, GUNA, VEERYA, VIPAKA, and PRABHAVA.

1. RASAS (tastes) are six—sweet, sour, salt, bitter, pungent, and astringent. Of these the first is more restorative than the second, the second more so than the third, and so on. The first three tastes (sweet, sour, and salt) are antagonistic to wind humour, and the other three to phlegmatic disorders. Astringent, bitter and

sweet tastes pacify biliary complaints; while salt, sour and pungent promote the secretion of bile.

The different tastes possess different properties, which are thus described:—

(*a*) MADHURA (sweet) taste has the property of increasing virility, promoting strength and secretion of milk in women, improving the eyesight, strengthening the body and germinating worms. It is beneficial to children, adults, the wounded, the bald and the feeble.

(*b*) AMLA (sour) promotes appetite and digestion, is cooling to perception but heating in effect; cures wind disorders, is laxative, but bad for semen. Habitual use of it causes amblyopia and other diseases.

(*c*) LAVANA (salt) is tonic, relaxes the bowels, deranges bile and phlegm, causes flaccidity, lowers the activity of the sexual functions and promotes perspiration. If continually taken it turns the hair white.

(*d*) KATU (pungent) is hot, destroys worms, diminishes the secretion of milk and dries the nose; promotes appetite and lessens the fat in the body. It improves the intelligence, but destroys strength and beauty.

(*e*) TIKTA (bitter) is cooling, alleviates thirst,

fainting, fever, and burning sensation; cures blood diseases, but causes derangement of wind. Taken in excess, it causes shooting pain in the head.

(*f*) KASHAYA (astringent) heals wounds, produces costiveness and softens the skin. If astringent substances are frequently taken they stiffen the body, swell the abdomen and cause pain in the heart.

2. GUNA (virtue) is the inherent property of a drug causing a particular effect when used either internally or externally.

India is a vast and fertile country, and has, as we have said before, the advantage of enjoying all the periodical seasons of the year. This circumstance makes it an encyclopædia of the VEGETABLE World. The ancient Aryans have taken the trouble to examine and study all the herbs that came under their observation, and classified them into Groups or *Ganas*. Charaka gives fifty groups of ten herbs each, which he thinks "are enough for the purposes of an ordinary physician," though at the same time he adds, that "the number of groups can be increased to any extent." Similarly, Sushruta has arranged 760 herbs in thirty-seven sets according to some common properties. Other writers have added to the list, which forms an interesting literature of the Materia Medica of India. They have also described the proper seasons for gathering the herbs, the period of their growth, when they possess their distinctive properties, the localities from which they should be collected, and the manner of treating them, extracting their active principles, and preserving them. Some of the groups mentioned by the Indian writers are given below :—

1. ANGAMARDAPRASHAMANA (Antispasmodic), as Vidarigandha (*Costus speciosus*).
2. ANULOMANA (Cathartic), as Haritaki (*Terminalia Chebula*).
3. ARSHOGNA (Hæmostatic), as Indrayava (*Wrightia antidysenterica*).
4. ARTAVOTPADAKA (Emmenagogue), as Jotishmati (*Cardiospermum Helicacabum*).
5. ASHMARIGHNA (Lithonlytic), as Gokshura (*Tribulus terrestris*).
6. AVRISHYA (Anaphrodisiac), as Bhoostrina (*Andropogon schœnanthus*).
7. BHEDANA (Purgative), as Katuki (*Picorrhiza kurroo*).
8. CHARDINIGRAHANA (Anemetic), as Dadima (*Punica granatum*).

9. CHHEDANA (Laxative), as Marichi (*Piper nigrum*).
10. DAHAPRASHAMANA (Antipyrotic), as Ushira (*Andropogon nardus*).
11. DAMBHA (Escharotic), as Bhallataka (*Semicarpus Anacardium*).
12. DEEPANEEYA (Stomachic), as Pippalimoola (*Piper longum*).
13. GARBHASRAVI (Ecbolic), as Grinjana (*Daucus Carota*).
14. GRAHI (Carminative and Exsiccative), as Jeeraka (*Cuminum Cyminum*).
15. HIKKANIGRAHA (Antisingultus), as Shathi (*Hedychium spicatum*).
16. JVARHARA (Antipyretic), as Peelu (*Salvadora indica*).
17. KAFAHARA (Antiphlegmagogue), as Bibheetaka (*Terminalia bellerica*).
18. KAFAKARA (Phlegmagogue), as Ikshu (*Saccharum officinarum*).
19. KANDUGHNA (Antipsoric), as Chandana (*Santalum album*).
20. KANDURA (Rubefacient), as Kapikachhu (*Mucuna pruriens*).
21. KANTHYA (Expectorant), as Brihati (*Solanum indicum*).

22. KARSHYAKARA (Antifat), as Gavedhu (*a kind of corn*).
23. KRIMIGHNA (Anthelmintic), as Vidanga (*Embelia ribes*).
24. KRIMIKRIT (Vermiparous), as Matha (*Phaseolus aconitifolius*).
25. KUSHTAGHNA (Antiscorbutic), as Haridra (*Curcuma Zedoaria*).
26. LALAGHNA (Antisialagogue), as Jatifala (*Myristica moschata*).
27. LALOTPADAKA (Sialagogue), as Akalakarabha (*Spilanthes oleracea*).
28. LEKHANA (Liquefacient), as Vacha (*Acorus Calamus*).
29. MADAKA (Inebrient), as Dhattura (*Datura Stramonium*).
30. MOOTRASANGRAHANA (Anuretic), as Pippalachhala (*Cortex Ficus religiosa*).
31. MOOTRAVIRECHANEEYA (Diuretic), as Kasha (*Poa cynosuroides*).
32. NIDRAHARA (Antihypnotic), as Shigrubeeja (*Moringa pterygosperma*).
33. NIDRAKARA (Hypnotic), as Kakajangha (*Capparis sepiaria*).
34. NIROMAKARA (Depilatory), as Rala (*Shorea robusta*).

35. PITTAHARA (Anticholeric), as Kamala (*Nelumbium speciosum*).

36. PITTAKARA (Cholagogue), as Tvak (*Cinnamomum Cassia*).

37. PRAJASTHAPANA (Anecbolic), as Vishnukranta (*Evolvulus hirsutus*).

38. PRAMATHI (Antiphysical), as Hingu (*Narthex Asafœtida*).

39. PRASAVAKA (Parturifacient), as Beejpura (*Citrus acida*).

40. PRATIVASA (Antibromic), as Karpoora (*Camphora officinarum*).

41. PURISHASANGRAHANA (Astringent), as Priyangu (*Panicum miliaceum*).

42. RASAYANA (Rejuvenescent), as Guggula (*Amyris pentaphylla*).

43. RECHANA (Hydragogue), as Trivrita (*Ipomœa Terpethum*).

44. ROHANA (Epulotic), as Tila (*Sesamum indicum*).

45. SAMMOHANA (Anæsthetic), as Madya (*Vina medicata*).

46. SAMSHODHANA (Emetic and Purgative), as Devadali (*Luffa echinata*).

47. SANKOCHANA (Constringent), as Mayofala (*Quercus infectoria*).

48. SANJEEVASTHAPANA (Restorative), as Jatamansi (*Nardostachys jatamansi*).
49. SHAMANEEYA (Calmative), as Amrita (*Cocculus cordifolius*).
50. SHEETAPRASHAMANA (Antalgide), as Agaru (*Aquilaria agallochum*).
51. SHIROVIRECHANA (Sternutatory), as Agastya (*Agati grandiflora*).
52. SHOFAKARA (Epispastic), as Snoohee (*Euphorbia Tirucalli*).
53. SHOOLPRASHAMANA (Anticolic), as Ajamoda (*Ptychotis ajowan*).
54. SHOTHAHARA (Discutient), as Arani (*Premna serratifolia*).
55. SHRAMAHARA (Refrigerant), as Ikshu (*Saccharum officinarum*).
56. SHRONITASTHAPANA (Styptic), as Kesara (*Crocus sativus*).
57. SHUKRAJANANA (Spermatopoietic), as Ksheerakakoli (*Hedysarum gangeticum*).
58. SHUKRALA (Tonic), as Rishabhaka (*Helekteres isora*).
59. SHUKRASHODHANA (Semen-improver), as Kushtha (*Saussurea lappa*).
60. SHVASAHARA (Antasthmatic), as Ela (*Amomum elettarum*).

61. SNEHOPAGA (Demulcent), as Vidari (*Batatas paniculata*).

62. SRANSANA (Drastic), as Rajataru (*Cassia fistula*).

63. STANYAJANANA (Galactagogue), as Shatapushpa (*Pimpinella Anisum*).

64. STHAULYAKARA (Fat-former), as Panasa (*Artocarpus integrifolia*).

65. SVARYA (Voice-improver), as Madhuka (*Glycyrrhiza glabra*).

66. SVEDOPAGA (Diaphoretic), as Punarnava (*Boerhaavia diffusa*).

67. TRISHNANIGRAHANA (Frigorific), as Dhana (*Coriandrum sativum*).

68. VAJEEKARA (Aphrodisiac), as Ashvagandha (*Physalis somnifera*).

69. VAMANA (Emetic), as Madana (*Randia dumetorum*).

70. VARNYA (Cosmetic), as Manjishtha (*Rubia cordifolia*).

71. VATAKARA (Flatus-producer), as Vallaka (*Dolichos sinensis*).

72. VEDANASTHAPANA (Anodyne), as Shireesha (*Mimosa Serissa*).

73. VISHA (Toxic), as Vatsanabha (*Aconitum Napellus*).

74. VISHAGHNA (Antitoxic), as Nirgundi (*Vitex Negundo*).

75. VYAYAYI (Sedative), as Bhanga (*Cannabis sativa*).

Agnivesha, a disciple of Charaka, enumerates no less than five hundred classes of medicinal agents, arranged according to their real or supposed virtues in curing diseases. A few classes have been selected from this and other sources and noted above. The chief notable feature in connection with the nomenclature of the Indian plants is, that in several cases their names are descriptive either of their character or property. A few instances of names descriptive of the prominent specific character of the herb may be given:—

(*a*) *Brachyramphus sonchifolius* is called Akhu-karni (rat-eared), as the leaves of the plant resemble the ears of a mouse.

(*b*) *Acorus Calamus* is called Ugra-gandha (strong-smelling), because it gives off a very pungent odour.

(*c*) *Clitoria Ternatea* is called Go-karni (cow-eared), from the supposed resemblance of the seeds to the ears of a cow.

(*d*) *Aconitum ferox* is called Vatsa-nabha

(calf's navel), because the root resembles in appearance the umbilical cord of a calf.

(*e*) *Sapindus emarginatus* is styled Bahuphena (very foamy), as, like soap, its berry produces much froth when agitated with water.

(*f*) *Ricinus communis* is called Chitrabeeja (spotted-seed), because of the seed being mottled with white, brown, or dark patches.

(*g*) *Mimosa sensitiva* is called Lajjalu (shy), from its leaves mimicking sensibility by folding themselves at the slightest touch.

(*h*) *Tribulus terrestris* is called Trikantaka (three-prickled), because its fruit is armed with three thorns.

(*i*) *Datura alba* is called Ghanta-pushpa (bell-flower), from the shape of its flowers.

(*j*) *Cassia fistula* is called Deergha-fala (long-pod), because its pod is cylindrical, about two feet in length, and one to one inch and a half in diameter.

The following are a few names descriptive of the inherent virtue of the herb :—

(*a*) *Amygdalus communis* is called Vatavairee

(wind-enemy), as it cures disorders of the wind.

(b) *Embelia ribes* is called Krimi-ghna (worm-killer), from its anthelmintic properties.

(c) *Cassia Tora* is known by the name of Dadru-ghna (itch-curing), as it is supposed to be very efficacious in curing the itch.

(d) *Coleus aromaticus* is named Pashana-bhedi (stone-breaker), as its juice is said to possess the property of dissolving stone.

(e) *Trianthema obcordata* is called Shotha-ghnee (intumescence-curing), from the use of its root in dispersing morbid swellings.

(f) *Ophelia Chiretta* is named Jvarantaka (fever-ending), for it is supposed to check fever.

(g) *Thevetia neriifolia* is called Pleeha-ghnee (spleen-curing), being credited with the power of curing splenic disorders.

(h) *Terminalia bellerica* goes by the name of Kasa-ghna (cough-curing), because it cures pulmonary catarrh.

(i) *Semicarpus Anacardium* is known as

Arushkara (eschar-causing), because when applied to a living part its juice gives rise to an eschar.

(*j*) *Cassia Absus* is called Lochana-hita (eye-benefactor), as its seeds are used as eye-salve to strengthen the sight.

3. VEERYA (power) is the third of the five properties innate in every medical substance, a knowledge of which is considered indispensable for a practical study of the Materia Medica. According to the influence of the sun or the moon a medicine is believed to be either hot or cold in power. It is therefore called "Ushna-veerya," heating, or "Sheeta-veerya," cooling. Hot agents cause giddiness, thirst, uneasiness, sweat and burning sensation; suppress cough and wind, but increase bile and promote digestion. Cold agents lessen bile and increase wind and phlegm, promote strength and pleasure and improve the blood. When a medicine capable of producing effects similar to the disease to be treated is administered, or, as the Homœopathists would put it, "*Similia similibus curantur*," it is on the principle that a patient suffering from the effects of inherent heat must be treated with a remedy apparently hot, but really cooling in its effects,

and *vice versa*, or otherwise the result would be disastrous. The general belief of the Hindoos in the hot and cold inherent qualities of medicines is fully shared by the Greek physician Galen, who teaches that, if a disease be hot or cold, a medicine with the opposite qualities is to be prescribed.

4. VIPAKA (consequence of action) is the change which a medicine undergoes in the organism under the influence of the internal heat. When a substance in the stomach is brought into contact with the digestive fire it is decomposed, and is sometimes recognisable in another form, with its medicinal activity greatly modified by the chemical changes that affect it. This converted state of the substance is called its Vipaka. The chemical effect on the six kinds of tastes is either sweet, sour, or pungent. The Vipaka of sweet, sour, and pungent agents remains unaltered as a general rule; that of a saline substance becomes sweet; and of astringent and bitter, pungent. To this (as to most other rules) there are exceptions. Rice, for instance, is sweet, but by the influence of the bodily temperature within, it turns sour. Chebulic myrobalans have an astringent taste, but by chemical action in the organism they become

sweet. A sweet Vipaka promotes phlegm but lessens wind and bile; a sour Vipaka increases bile but decreases wind and phlegm; while a Vipaka that is pungent gives rise to disorders of wind and subdues those of phlegm and bile. Native Pharmacodynamics treat of the changes which each medicinal agent undergoes in the organism. In determining the property of an agent and the chemical changes that affect it, the ancients have ascertained which of the five constituent elements—ether, wind, fire, water, and earth—is predominant in its formation. The five elements have been characterised by their respective qualities of lightness, dryness, sharpness, unctuousness and heaviness. It may be noted here, by way of parenthesis, that this elemental theory precisely accords with that of Plato, Hippocrates, and Pythagoras, though the first two do not seem to recognise ether as an elemental constituent. To determine the proportion of the several elements in the formation of a medicinal drug, and to describe the subsequent changes it undergoes in the living economy, presupposes some knowledge, on the part of the old Indian writers, of chemical analysis and the process of decomposition.

The therapeutic effect of a medicinal agent is regulated, not by the nature of its inherent taste, but by that of the taste of its Vipaka.

5. PRABHAVA (inherent nature) is the peculiar active force residing in a drug. There are certain drugs whose taste, property, power, and consequence of action are analogous, and yet the effects produced by them are quite dissimilar. For example, Madhusrava ( *Bassia latifolia* ) and Draksha (*Vitis vinifera*) are similar in taste, both being sweet; similar in property, both being heavy; similar in power, both being cold; and similar in consequence of gastric action, both remaining sweet in their Vipaka, and yet the physiological effect of the former is costive and that of the latter laxative. This inherent peculiarity of the drugs is called their Prabhava. In like manner, Chitraka (*Plumbago zeylanica*) and Danti (*Croton polyandrum*) are both pungent to taste, light in property, hot in power, and pungent in consequence of gastric action. But Chitraka promotes digestion, while Danti operates as a powerful purgative. Certain substances show their Prabhava independently of the four conditions emunerated above. For instance, a herb called Sahadevi (*Vernonia*

*cinerea*), if procured in a prescribed manner and tied on the head, is said to cure intermittent fever, though as an ordinary medicine, when administered internally, it is an alterative and a bitter tonic, and its juice when applied externally is supposed to cure leprosy and chronic skin-diseases. It is under this belief that persons acquainted with the Prabhava, or efficacy of certain objects, as fruits or stones, wear them on their bodies as prophylactics against certain diseases. The ascetics of India, who prefer to be aloof from society and pass their time in the solitude of the jungles, are said to be familiar with the wonderful properties of rare drugs, which go not only to keep their bodies and souls together, but to prolong their lives to a considerable extent. Their knowledge of the Prabhava of the different herbs, combined with the practice of regulating their breathing, is supposed to give them a longevity quite beyond our comprehension. This knowledge is handed down from teacher to pupil, and forms no small volume of the unwritten and traditional lore on the subject of the nature and properties of the Indian curative agents.

The Materia Medica of India is acknowledged

on all hands to be very voluminous. But the most noticeable feature in connection with this particular branch of Aryan medical science is, that unlike other Aryan sciences it has been up to a certain period a progressive one. Each successive writer, after a patient and careful investigation, appears to have added new drugs to the existing list, and to have thus conferred a lasting benefit on mankind. Some of the writers emphatically assert that all the curative agents mentioned in their treatises have been thoroughly tested and recommended after a long practical experience. Each writer has of course his own method of treating the subject. We have already referred to the classification of Agnivesha and of Sushruta. The latter, in the 39th chapter of his standard work, has arranged the drugs into classes according to their power of curing certain diseases, prescribing from ten to twenty-five remedies for each disease. He strongly recommends that physicians should be able to identify the various remedial agents they have to deal with. They should personally go to the jungles, and with the help of shepherds, graziers, ascetics, travellers, and others familiar with the forests, gather the herbs when they are in flower, taking care to avoid those injured by insects, or growing on situations containing nests of white ants, or where bodies have been burnt or buried, or from ground in which there is much salt. We have also referred to the classification of Charaka, based on the properties of various substances. Vagbhata, in the 15th chapter of his popular work, has followed Sushruta's method, but the concise way of his description has a charm of its own. The method adopted by the author of "Dhanvantari Nighanta" is much the same as followed by Charaka, with this difference, that while the latter mentions one drug in the treatment of several diseases, the list of the former is free from such a repetition. The work is of great antiquity, but the name of the compiler is not known. Some ascribe the authorship to Dhanvantari, the Father of Indian Medicine. But this cannot be correct. For in the prologue of his work the writer offers his salutations to "the Divine Dhanvantari adored alike by gods and demons." In his elaborate work he has treated of 373 drugs, exclusive of minerals.

The next important writer on medicinal herbs is Bhava Mishra, son of Lataka Mishra, to whom a reference has already been made in previous pages. He has given the names and properties of about 150 drugs more than are found in "Dhanvantari Nighanta," such as Ahiphena (opium), Khakhas (poppy seeds), Kasumba (safflower), Methica (fenugreek), Vatavairi (almond), etc.

Bhava Mishra is followed by Raja Madanapala, whose work called "Madana-Vinoda" is a second edition, as it were, of the "Bhavaprakasha." He seems, however, to have augmented the list of Indian plants by some new names, among which might be mentioned "Akakarabha" (pellitory), "Anjira" (fig), "Pistam" (pistachio nut), "Haridruma" (gambier), etc.

Just about his time there flourished a learned physician named Narahari, son of Chandeshvara, an inhabitant of Sinhpur in Cashmere. He wrote an excellent work called "Abhidhana Chudamani" or "Raja Nighanta" (Royal Dictionary of Medicine). The work was composed under the patronage of the King of Cashmere at the time, and therefore no pains seem to have been spared to make it as useful and interesting as possible. According to some writers, Narahari lived in the seventh century after Christ, though the exact time of his birth is not known. His work is a glossary of medicinal substances with specifications of their virtues. He also describes the properties of different kinds of soil; the nature of soils suitable for the cultivation of various medicinal plants; varieties of trees, cereals, oils, vegetables, roots, leaves, flowers and fruits; properties of fresh and salt waters; and gives, besides, a mine of useful information. The work is very elaborate, and is much valued by Indian practitioners. The order observed by this writer in arranging the drugs differs from that of his predecessors. He classifies the herbs into creepers, plants, trees and grasses, and describes how each part of them is to be used medicinally. This writer makes mention of about a hundred new medicines not to be met with in the works of his predecessors. The most important of them are: KANDURA (*Gyrardinia heterophylla*), BRAHMADANDI (*Tricholapis glaberrima*), JINJHIRA (*Triumfetta angulata*), RUDANTI (*Cressa cretica*), and RUDRAKSHA (*Elæocarpus ganitrus*).

Shodhala, who came after Narahari, wrote a treatise on Materia Medica bearing his name. He was a Gujarati Brahman by caste, his father being a physician named Nandana. His work is chiefly based on the "Dhanvantari Nighanta," to which he has added about eighty drugs as the result of actual investigation carried on in the forests such as Mamanjaka (*Hipian orientali*), Jhullapushpa (*Byophytum sensitivum*), Keetamari (*Aristolochia bracteata*); Utkantaka (*Echinops echinatus*), Bhringaraja (*Eclipta alba*), etc.

Vaidya Moreshvara of Ahmednagar, in the early part of the seventeenth century incorporated in his "Vaidyamrita" some Persian drugs, as Isphgul (*Plantago Ispaghula*) and others.

In the beginning of the eighteenth century, a well-known physician of Benares composed a large work called "Atankatimirabhaskara," an important work on the Indian healing art. In the chapter on Materia Medica, he has not only availed himself of the labours of all who had gone before him, but has thrown a new light on some of them. Tea is one of the few new drugs he has embodied in his work. His great-grandson, Vaidya Sohamji, was one of the most scholarly and celebrated physicians in Northern India. He died very recently.

About the middle of the present century, that is to say in 1867, Pandit Vishnu Vasudev Godbole published his "Nighantaratnakara." It is a very popular work, as it contains an epitome of all the previous treatises on Materia Medica, supplemented by about fifty new herbs not referred to by the older writers. Among the new names we find Elivaka (aloes), Anannasa (pineapple), Peruka (guava), Tamakhu (tobacco), Pudina (mint), Medica (henna), Sitaphala (custard apple), etc.

The virtues of the Indian drugs were known not only in the country of their birth, but in other countries as well. Some five centuries before Christ, Hippocrates in his Materia Medica recommends several Indian plants mentioned in Sanskrit works of much anterior date, as for instance *Sesamum indicum* (tila), *Nardostachys jatamansi* (jatamansi), *Boswellia thurifera* (kunduru), *Zingiber officinale* (shringavera), *Piper nigrum* (marichi), etc. In the first century of the Christian era Dioscorides, a Greek physician, thoroughly investigated the medicinal virtues of many Indian plants which were then taken to the market of Europe, and incorporated in his extant book on Materia Medica, which for many ages was received as a standard work. In the second century, Claudius Galen, to whose writings modern European science is indebted for many useful discoveries, published his famous work, the leading opinions in which as to hot and cold medicines were borrowed from India, where they still prevail. Ætius, a physician of Mesopotamia, who flourished in the fifth century, and whose works on the diseases of women are still extant in Greek, mentions some drugs, as Indian nuts, sandal-wood, cocoa-nuts, and other products of India. The Ægian physician, Paulus Ægineta, who is said to have first noticed the cathartic quality of Rhubarb, and who lived in the seventh century, refers to certain Indian herbs in his work. In the eighth century, and probably in the century following, the natives of India practised as physicians in Baghdad, and employed many Indian drugs in their practice.

We find from the books written by Arabian and European travellers of bygone days, that about 600 A.C., the Arabs, who were the most forward and enterprising nation of their time, used to bring various articles of merchandise to India from their own country, and from countries lying on the east coast of Africa, and took with them from the Malabar coast in Southern India spices and medicinal drugs, and so spread a knowledge of these articles in the adjoining countries of Europe. This state of things continued for a long time, while the Medical Science of India was in its heyday of glory. Every important town could then boast of one or more medical schools, the pupils at which used to accompany their teachers to the jungles to identify for themselves the various drugs mentioned in their books. The physicians, in their laborious researches, were very liberally encouraged by the ruling chiefs—great and small—in all parts of the country. So long as they continued to receive encouragement from the kings, the science prospered and flourished. Its decline dates from the Mahomedan invasions in the tenth century. The minds of both princes and people were distracted by these foreign intruders. They were chiefly engrossed in taking measures for opposing the invaders. It was only natural that during such a state of unrest and disorder, the native Vaidyas should slacken their zeal for making further investigations in the Indian flora, for want of encouragement. Far from being able to follow up the practical part of their study, they had to rest content with the theoretical knowledge imparted by their books, and to depend on ordinary grocers for the supply of drugs required for their nostrums. When the Mahomedan power was firmly established in

India, the Indian medicine received a rude shock. For the Mahomedans brought with them their own physicians, called Hakims, who followed in their practice the Ionian (Greek) system of medicine, generally termed "Yunani." Under Imperial patronage the Hakims began to prosper at the expense of the Vaidyas. But even at the Mahomedan Courts the Vaidyas are recorded to have cured many intractable diseases, which had baffled the skill of their foreign rivals. It is evident that during the time of the Mahomedan rule there were introduced into India some new drugs from Arabia, Persia, and Afghanistan. Opium, for instance, appears to be a native of Western Asia. It was first imported into this country from Arabia. Its spread in India is synchronous with the advent of the Mahomedans who had adopted it as a suitable substitute for fermented liquors, which their religion discountenances. Sharngdhara and Vagbhata refer to the medicinal use of this article, which they call "Ahi-phena" or snake-foam, believing it to be inspissated saliva of snakes, probably from the symptoms of opium poison resembling those produced by the venom of snakes. It is used in diarrhœa, chronic dysentery, for allaying pain, and producing sleep. The European doctors seem to have learnt the therapeutic use of opium from Indian practitioners, though Scribonius Largus has noticed *Opium* early in the first century. Some more drugs which happened to be introduced into India during the Mahomedan rule are :—

ALU (*Prunus bocariensis*) is used in bilious affections and fevers.

BADIAN (*Illicium anisatum*) is a Persian drug, and its oil is applied to the joints in rheumatism.

BANAFSHA (*Viola odorata*) is employed in bilious affections and constipation.

GAOZABAN (*Onosma bracteatum*) is used in leprosy, hypochondriasis, and syphilis.

GUL-E-DAUDI (*Chrysanthemum Roxburghii*) is prescribed as a demulcent in gonorrhœa.

KERBA (*Panitis succinifer*) is antispasmodic and stimulant.

KHARJURA (*Phœnix dactylifera*) is nutritive and used as desert.

The Mussulman rule was supplanted by the English, whose power was firmly established in India in the eighteenth century. The English brought with them their own doctors, who prescribed European medicines, before which the indigenous drugs gradually gave way. Hospitals and dispensaries on Western models multiplied, and the use of Western medicines was encouraged in all parts of the country. Native medicines came to be discarded in favour of ready-made preparations imported from Europe. This was a serious blow to Indian pharmacy. But Europe is simply paying back the debt it owed to India, because its Materia Medica includes many curative agents of Indian product, such as :—

*Aconitum heterophyllum* (Ativisha).
*Allium Cepa* (Palandu).
*Acacia Catechu* (Kadara).
*Alhagi maurorum* (Yavasa).
*Alstonia scholaris* (Saptaparna).
*Amomum elettarum* (Ela).
*Andropogon nardus* (Ushira).
*Andropogon Schœnanthus* (Katurina).
*Artemisia sternutatoria* (Agnidamani).
*Berberis Lycium* (Daruharidra).
*Butea frondosa* (Palasha).
*Cassia lanceolata* (Sonamukhi).
*Cucumis Colocynthis* (Indravaruni).
*Datura alba*, *Datura niger*, etc. (Dhattura).
*Justicia Adhatoda* (Atarusha).
*Luffa amara* (Katukoshtaki).

*Linum usitatissimum* (Atasi).

*Mallotus Philippiensis* (Kapillaka).

*Myrica sapida* (Katfala).

*Ophelia Chiretta* and *Ophelia angustifolia* (Kirata).

*Pimpinella Anisum* (Shatapushpa).

*Pongamia glabra* (Karanja).

*Ptychotis ajowan* (Ajamoda).

*Ricinus communis* (Eranda).

*Salvinia cucullata* (Undurkarnika).

*Santalum album* and *Santalum flavum* (Chandana).

*Shorea robusta* (Ajakarna).

*Strychnos potatorum*, *Strychnos nux vomica*, etc. (Katakafala).

*Tinospora cordifolia* (Gaduchi).

*Valeriana Hardwickii* (Tagara).

*Wrightia antidysenterica* (Indrayava).

The Hindoos from an early date have derived simple medicines from the ANIMAL Kingdom. Their number is very large. A few may be noted here:—

ASTHI (bone) of a goat reduced to ashes, and formed into an ointment with other ingredients, is used for curing fistulæ. Cuttlefish bones are also used medicinally.

DANTA (tooth) of the elephant is prescribed in leucorrhœa.

DUGDHA (milk) is nutritive and vitalising. Human milk is light and strengthening and much used in eye diseases. Cow's milk increases the secretion of semen. Buffalo's milk induces sleep when taken in large quantities. Goat's milk is sweet and light, and is recommended in phthisis and blood diseases. Sheep's milk is hot, and is believed to promote the growth of hair. Elephant's milk is used in eye diseases, and mare's milk in rheumatism. Ass's milk is saltish, and is supposed to relieve cough in children. Camel's milk is laxative, and is used in dropsy, asthma, and scrofulous diseases. The properties of milk are said to vary according to the colour of the animal and the qualities of the pasture. The chief preparations of milk are Dadhi (curds), a favourite remedy for diarrhœa; Takra (whey), which is refrigerant; Navanita (butter), used in constipation; Ghrita (clarified butter), is tonic, emollient, and cooling; Santanika (cream) is strengthening.

GARALA (venom) of snakes is used in dropsy.

TVAK (skin), which a snake periodically casts

off, is said to be an insecticide, and possesses several healing properties.

JALA (cobweb) of a house-spider is a useful application for stopping hæmorrhage.

JALUKA (leeches) are applied for bloodletting.

JEEVATA (living creatures), such as Matkuna (bed-bug), is alleged to cure quartan fever if swallowed. Similarly, a fly is swallowed to cause vomiting.

KESHA (hair) of a man when burnt and reduced to ashes, is applied to sores on the skin. The burning of hair is also resorted to for driving away serpents.

LAKSHA (lac) is used in menorrhagia.

MADA (the secretion that flows from an elephant's temples when in rut) has its medicinal use in exciting sexual desire. Similarly, Kasturi (musk) is used in hysterical disorders and other nervous affections.

MADHU (honey) is demulcent and laxative, and is used both internally and externally. Hindoo writers describe eight kinds of honey, viz.:—Makshika, secreted by big tawny bees, and considered to be the best, is recommended in jaundice. Bhramara is white and cures scurvy. Kshaudra is secreted by small tawny bees and is used in gonorrhœa. Pautika is secreted by tiny black bees, is hot in property, and cures stricture of urethra. Chhatra can be had on the Himalayas where the honeycomb is found in the shape of a Chhatra or umbrella, and is employed in expelling worms. Ardhya is found in Malwa, and is said to be very beneficial in eye diseases. Audalaka is obtained from the ant-hills, and is good for the voice. Dala* is the juice exuded from certain kinds of flowers and collected on the leaves. It is credited with the property of curing nausea. Honey of a particular kind of rhododendron is poisonous.

MADHUJAMA (wax) is used in cerates and ointments. It is also given in the form of emulsion in diarrhœa and dysentery.

MANSA (flesh) of a goat fried in oil is used in rheumatism. An essence of dove's flesh is prescribed in paralysis.

MEDAS (fat) of camel or hyena is considered a valuable local remedy for gouty joints.

MOOTRA (urine) would appear to be a very useful agent, according to the Hindoos, and has

* This substance is a purely vegetable product, and is mentioned here because the Hindoo writers have generally classified it with the other varieties of honey.

a very wide application. Cow's urine is used both internally and externally. It is prescribed in colic and many other diseases. Goat's urine is used in jaundice, buffalo's in piles and elephant's in blood diseases. The renal secretion of the horse is prescribed for killing worms, of the ass for consumption and insanity, and of the camel for the cure of ringworm. Human urine is recommended for cough and eye diseases, and urine of a castrated bullock in cases of anæmia and dysentery. Urine should, as a rule, be obtained from the female; but in the case of the horse, the ass, the camel and the elephant, that obtained from the male is generally preferred.

MUKTA (pearls) are taken in a powdered state for impotency and consumption.

NAKHA (nail) of a man is used in cases of wounds, and horse's hoof for fumigation and intermittent fevers.

PICHHA (feather) of a peacock is said to cure hiccough. It is also believed that snake poison will not affect one wearing a ring made of copper extracted from peacock's feathers.

PITTA (bile) of fish and other aquatic creatures is helpful in cases of fever and eye diseases.

PRAVALA (coral) is beneficial in cough.

PURISHA (dung) of a cow is applied to parts of the skin that may be inflamed or discoloured. It is occasionally given internally. In India it is used for plastering the walls, and is spread on floors under the impression that it possesses disinfecting properties. Elephant's fimus is said to cure leprosy. Droppings of a domestic cock are considered beneficial in colic, and those of a goat in cutaneous diseases.

SHANKHA (conch) relieves colic and flatulence.

SHRINGA (horn) of a stag has various medicinal uses. Made into a paste, it is applied to sprains, contusions and fissures, and to the forehead in headache.

VARATAKA (cowry) is recommended for enlarged spleen.

The MINERALS used in medicine by the Hindoos include Metals, *Rasas*, Salts, Precious Stones, Clay, etc.

The Metals employed by the Aryan physicians are divided into two classes — principal and secondary. The principal metals or Dhatus are seven, namely: — SUVARNA (gold), RAUPYA silver), TAMRA (copper), BANGA (tin), SISAKA (lead), YASHADA (zinc) and LOHA (iron). The "secondary metals" (substances containing any of the principal metals or their compounds) possess the properties of such metals, though in a lesser degree, and are also seven, viz. :—Suvarnamakshika (yellow pyrites), Taramakshika (white pyrites), Tuttha (sulphate of copper), Kansya (brass), Reeti (calcined zinc), Sindura (red oxide of lead) and Shilajita (bitumen).

PARADA (mercury) is treated under the name of RASA (pleasure), as its presence in the composition of medicines is supposed to afford great satisfaction to the Vaidyas. It is called the principal Rasa, as distinguished from the Uparasas or secondary Rasas, which are Gandhaka (sulphur), Hingula (red sulphide of mercury), Abhraka (mica), Manasshila (bisulphide of arsenic), Talaka (tersulphide of arsenic), Srotanjana (sulphide of lead), Tankana (borax), Rajavarta (lapis lazuli), Chumbaka (loadstone), Sfatika (alum), Kasisa (sulphate of iron), Rasaka (carbonate of zinc) and Bodara (litharge).

The precious stones (RATNAS) mentioned in Materia Medica are also divided into two classes —principal and secondary. The principal gems are nine, their names being: Heera (diamond), Padmaraga (ruby), Nila (sapphire), Garutmat (emerald), Pushparaga (topaz), Gomeda (onyx),

Vaidurya (cat's eye), Mauktika (pearl) and Pravala (coral). The last two belong to the Animal Kingdom, but are referred to here as being included in the "nine gems." Among the secondary stones may be mentioned Suryakanta (sun-stone), Chandra-kanta (moon-stone—a gem supposed to be formed of the congealed rays of the moon), Sphatika (crystal), Haritshyama (turquoise), Kacha (glass), and some others.

Certain kinds of sand and clay are in common use as healing agents, such as Khatika (carbonate of calcium), Kardama (hydrous silicate of alumina), Gopichandana (silicate of alumina), Sikata (silica), etc.

The principal salts included in the Hindoo Pharmacopœia are Navasadara (chloride of ammonium), Sindhava (chloride of sodium), Pamshujakshara (carbonate of potassium), Yavakshara (carbonate of soda), and Suryakshara (nitrate of potash).

Besides the compounds already enumerated, the Hindoos have for ages past employed Jangala (subacetate of copper), Mandura (hydrated oxide of iron), Pashanabheda (carbonate of iron and lime), Yashadapushpa (oxide of zinc), Rasasindura (sulphide of mercury), Rasakarpura (corrosive sublimate), Shankhavisha (arsenious acid), and several other metallic preparations.

The metals have been recognised as remedies by the Hindoos from prehistoric times. Vegetable drugs are universally used as therapeutic agents. But it is to be remembered that preparations of vegetable substances do not keep well. The ancient Aryans seem to have ascertained by practical experience that vegetable drugs, as a general rule, become inert after a year; powders preserve their strength for two months, pills and tinctures for a year, and oleaginous preparations sixteen months. Under the circumstances, the Aryans have, it is alleged, discovered retentive and lasting medicines, which, far from becoming weakened in effect under the influence of time, increase in strength in proportion to their age. They have described the method of transferring the properties of vegetable cures to certain metals, which intensify their efficacy, and retain it a long time. The metals are subjected to various processes of purification, oxidation, etc., before they can be administered as medicines for various diseases. These Compounds or *Bhasmas*, as they are called, are supposed to be infinitely more effective than the vegetable drugs, and are always given in small doses. The number of physicians using mineral remedies is not large; for the general belief is that the metals, if not carefully and properly prepared, do more harm than good. Only those who are experts in this practice inspire confidence in their patients. The literature on metallic remedies is very voluminous among the Hindoos.

As has already been said, the metals before being calcined must in the first place be purified. Different modes are prescribed for purifying different metals. One of the simple processes of purifying gold is to manufacture the metal into thin plates; make these red-hot, and then dip them into sweet oil; again heat the plates red; plunge them into whey; heat them a third time; cool them in cow's urine and sour gruel; repeat this process seven times, and lastly dip the red-hot metal into Kulatha (a kind of vetch). The metal then becomes pure and free from all deleterious matter. It is then subjected to the process of "killing" or oxidation, with the object of reducing it to *bhasama.* Of the many processes described, the following may be taken as an example:—Let the purified gold be melted in a crucible with one-sixteenth part of its quantity of lead; triturate the mass in lemon juice and make it into a ball; then coat it with powdered sulphur; put the bolus in an earthen pot, and cover it with another vessel of the same size; cement the two together with a layer of white clay, and place them in the midst of fire made of twenty cow-dung cakes. When the fire has completely burnt out, take out the mass from the crucible. Repeat the process for seven consecutive days, and then the metal becomes calcined, and can easily be reduced to powder. This gold "calx" is said to be a good tonic, and is supposed to cure nearly all diseases. It is said to remove the effects of old age, and to restore the vigour of manhood, to sharpen the memory, improve the voice and colour of the body, and promote strength. It is stimulant and aphrodisiac.

Silver *bhasma* is prepared in very much the same way, and is highly recommended in sexual weakness and obesity.

The process of purifying copper is similar to that adopted for gold. When purified, boil the thin plates of the metal for three days in lemon juice, and incorporate it with one-fourth of its quantity of quicksilver. Then the mass, mixed with two parts of sulphur, is to be moulded into a ball and covered over with a layer of Punarnava (*Boerhaavia diffusa*) about one inch thick. Place this in an earthen pot, and roast it in Valukayantra (an apparatus, *see* Plate IV.) for twelve hours. After taking it off the fire, put the mass into the hollowed root of Shurana (*Arum Colocasia*), cover it with a coating of dung and clay, and expose it to heat. This makes the metal fit for reduction to powder, which is used in cases of enlargement of the liver and spleen, gout and rheumatism.

In order to purify tin, first melt and then soak it three times in succession in oil, whey, Kanjika (sour gruel), cow's urine, and lastly in the juice of Arka (*Calotropis gigantea*). Melt the purified metal in an earthen crucible, adding to it the powder of tamarind and banyan tree barks in the proportion of four to one. Stir and rub them with an iron ladle. Mix with the powder an equal quantity of Talaka and triturate in an acid juice. Expose it to fire; again add one-tenth of its quantity of Talaka, and again rub it and put it over a fire. Repeat the process ten times, or until the metal is reduced to *bhasma*, which is said to be a good remedy for painful micturition and other urinary disorders. It is also credited with the power of curing gonorrhœa, jaundice and obesity.

The process of purifying lead is similar to that of tin. In order to make it fit for medicinal use cover the mass of purified lead with a coating of Manasshila macerated and rubbed up in betel-leaf juice, put it on the fire and repeat the process thirty times, when the metal is converted into *bhasma*, which is a vermifuge, and is recommended for chronic diarrhœa.

The method of preparing zinc *bhasma* is the same as that employed for tin. The medicine is a nervine tonic and is used in cholera and epilepsy.

Iron is purified by exposing it to the fire of a furnace and quenching it three times successively in oil, Kanjika, cow's urine and Kulatha. Then to twelve parts of the metal add one of Hingula and triturate in the juice of Ghritakumari (Indian aloes) for six hours; expose it to the fire of Gajaputa (a square hole dug in the earth about two feet deep and two feet wide filled in with cow-dung cakes, in the midst of which the earthen vessel containing the metal to be roasted is put). After repeating the process seven times, iron can easily be reduced to *bhasma*, and is then prescribed for hectic fever, anæmia, dropsy and brain diseases.

Precious stones, like metals, must be subjected to the processes of purification and calcination before they can be used therapeutically. Diamond *bhasma* is credited with many wonderful properties. To purify a diamond it is placed in the hollow of a Vyaghri (*Solanum indicum*) root covered over with buffalo's dung. This is kept over a fire during the whole night, and quenched in horse's urine in the morning. The operation is repeated for seven days. Thus purified it is heated and cooled in a decoction made of asafœtida, bay salt and gruel. Go the same round for twenty-one days, and diamond *bhasma* is prepared. It is said to improve the colour of the body and cure many diseases. It will be interesting to note here that the Hindoos distinguish four kinds of diamonds, differing from each other in appearance and property, called Brahma, Kshatriya, Vaishya and Shudra, names derived from the castes into which the Hindoos are arranged. The Brahma diamond is clear white, the Kshatriya of reddish colour, the Vaishya is yellowish and the Shudra diamond is of a smoky colour. A diamond which is perfectly symmetrical, sharp-edged, lustrous, big, and without a stain is called a Purusha or a male diamond. It is considered the best as far as its medicinal use goes; it restores vigour and can be prescribed to all with advantage. That which exhibits stains and cracks, and which is hexangular, is distinguished as a female diamond, and its *bhasma* is beneficial to females only. Diamonds that are long and triangular are neuter and are considered powerless.

But of all the minerals Mercury is recognised as the most important by the Indian physicians. Marvellous powers are attributed to it. It was no doubt known to the Romans and Arabs who employed it externally, but the Hindoos seem to be the only people who prescribed it internally. It is found in many parts of India, and was known to its people from a very early date. Being a volatile substance it is unmanageable for purification and 'killing' without a good deal of care and patience. But if once brought under subjection, it proves, say the Hindoos, an invaluable medicine for curing some of the most obstinate diseases. The Vaidyas use various contrivances called Yantras for the preparation of mercurial compounds. These Yantras are known by the names of Damaru, Urdhvanalika, Valuka, Bhudhara, Dola, Patala, Nabhi and many others. For diagrams of some of them, *see* Plates I.–VI. Like other minerals mercury should be purified for medicinal purposes. This can be done in various ways. Sharngdhara gives the following directions:—Rub mustard seeds and garlic together until reduced to the consistency of mud. Make two small cups of the mass. Put mercury in one of them, lute the other over it, and dry it in the sun. Tie the vessel in a piece of cloth, suspend it for three days in an earthen pot filled with Kanjika and place it on the fire. Take out the mercury from the vessel and rub it for one day in Ghritakumari juice, another day in a decoction of Chitraka (*Plumbago zeylanica*), for the third day in that of Kakamachi (*Cocculus indica*), and the fourth day in that of the three Myrobalans. Wash the mass in Kanjika, to separate the mercury from it. Again put the quicksilver in a mortar containing half its quantity of bay salt, and triturate it for a whole day. Add an equal quantity of mustard seed, and rub the mass in a rice-husk decoction. Repeat the whole process with garlic and then with sal-am-

Plate I.

1. Ardhvakhalva yantra.

2. Adhahpatan yantra.

3. Baka yantra.

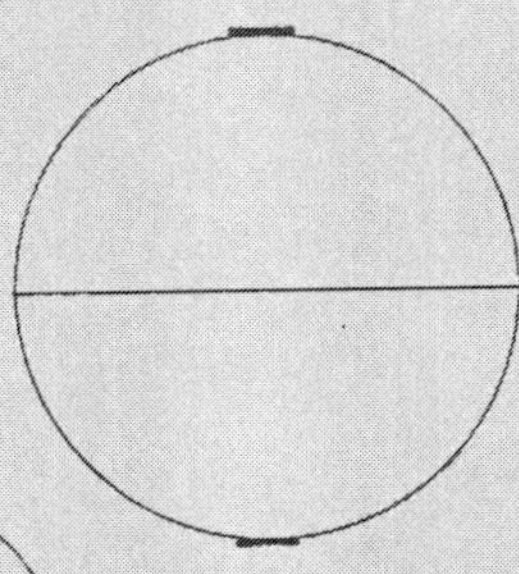

6. Samputa yantra.

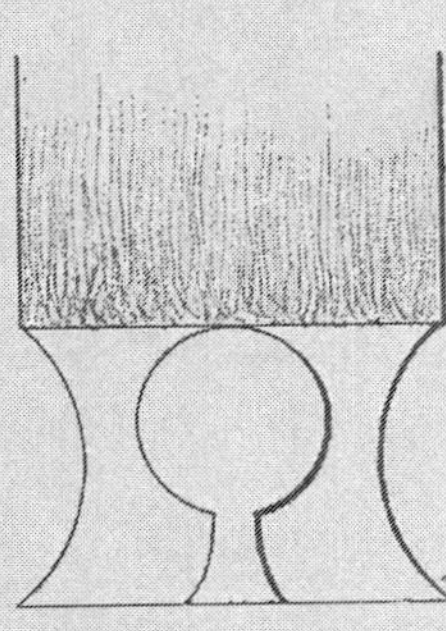

4. Bhudhara yantra.

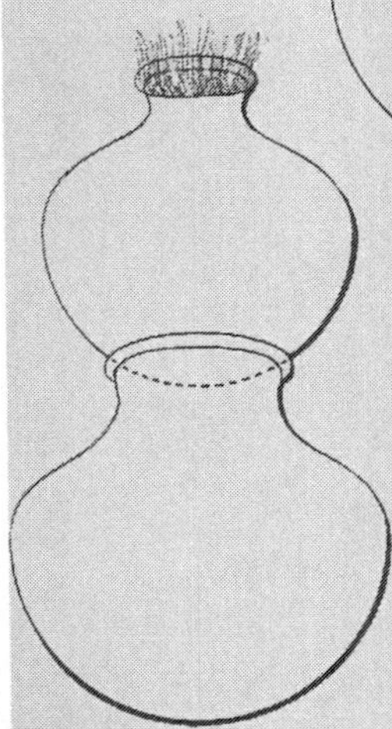

5. Somanala yantra.

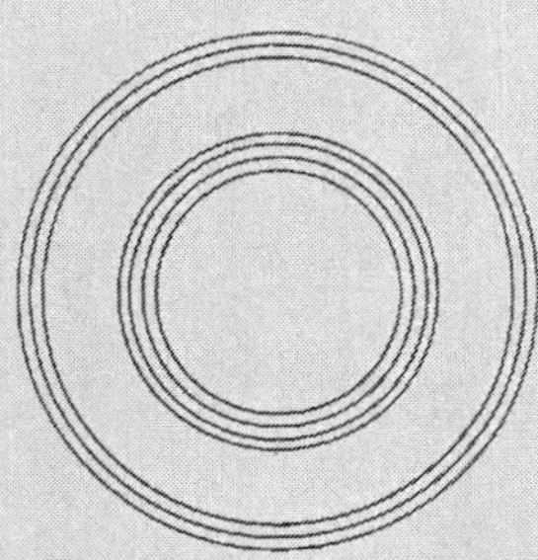

7. Chakra yantra.

F. Huth, Lith[r] Edin[r]

Plate II.

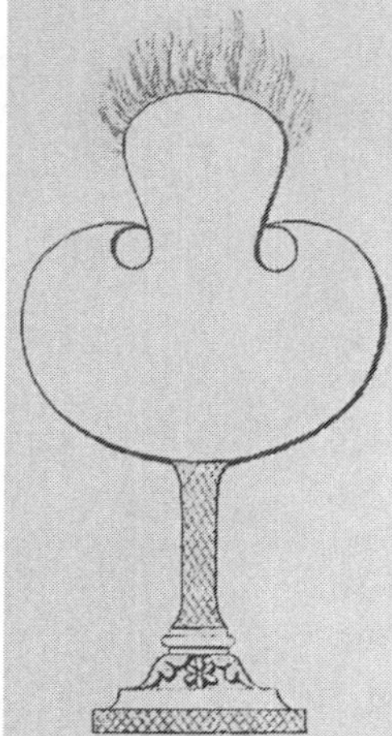

8. Dipika yantra.

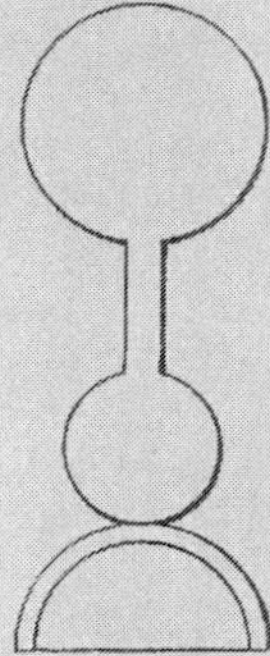

9. Damaru yantra.

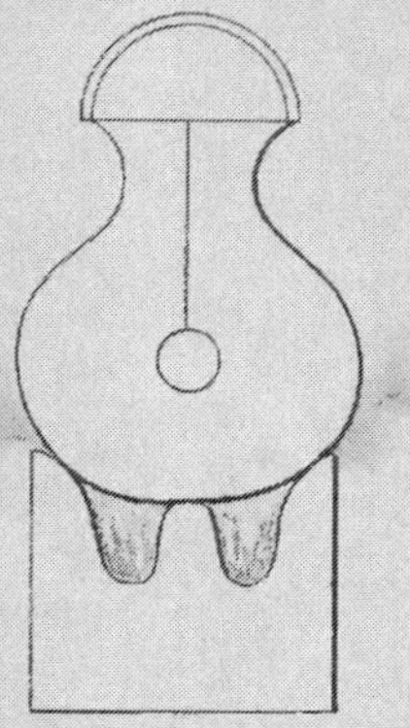

10. Dola yantra.

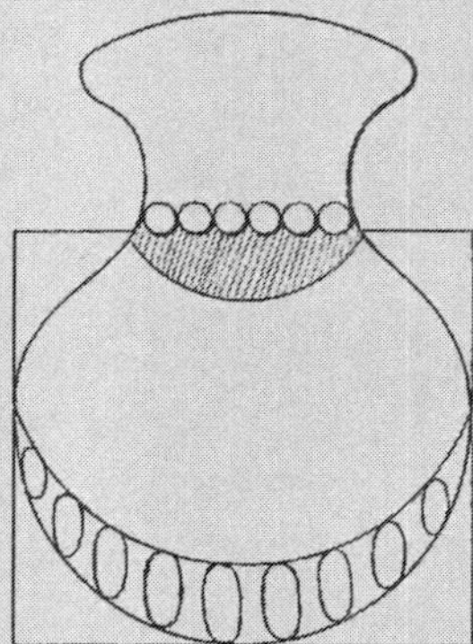

13. Ishtika yantra.

11. Dheki yantra.

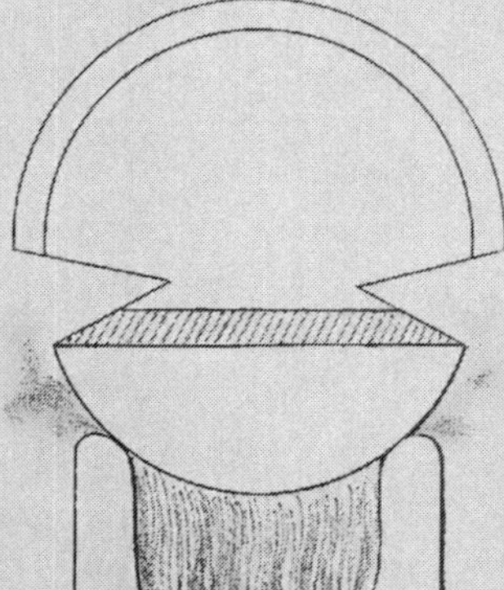

12. Dhupa yantra.

F. Huth, Lith[r] Edin[r]

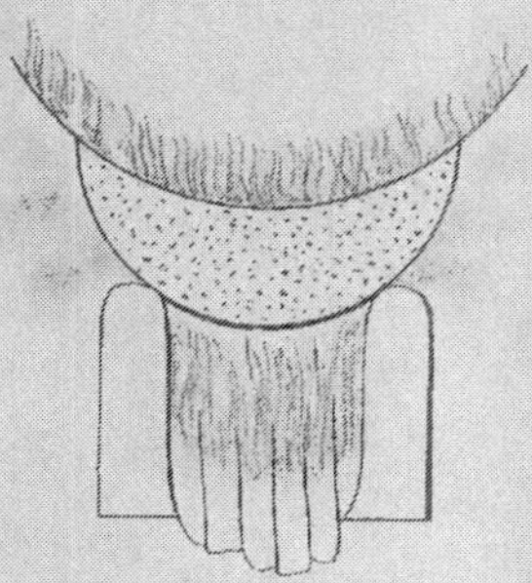

14. Hansapaka yantra.

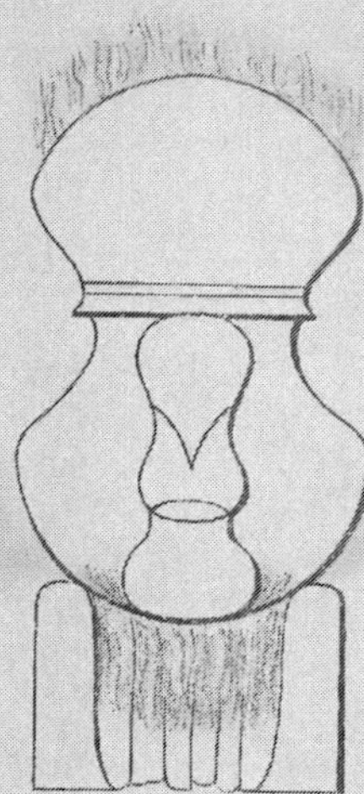

15. Jarana yantra.

18. Koshti yantra No. I.

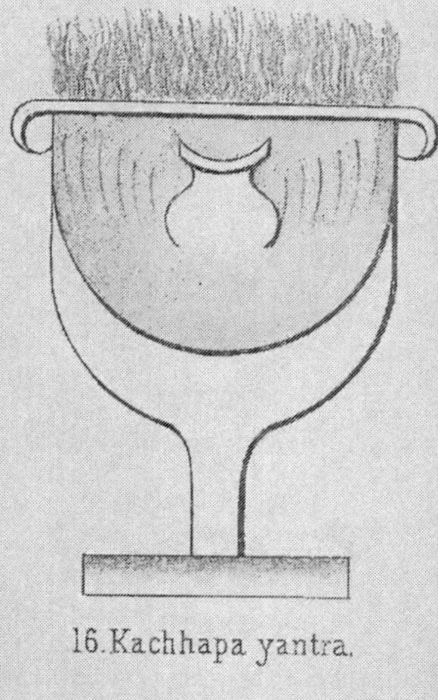

16. Kachhapa yantra.

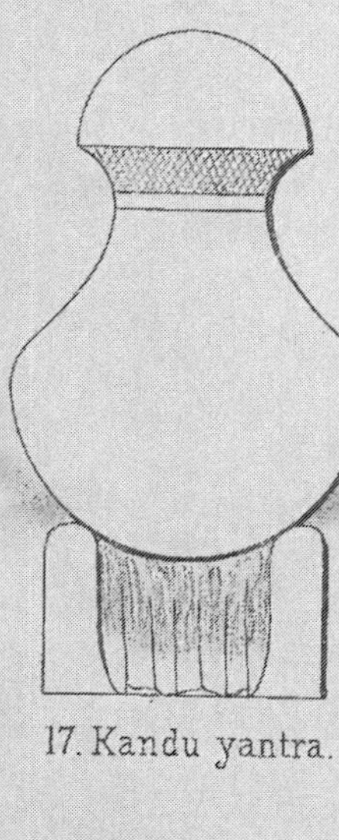

17. Kandu yantra.

F Huth, Lith[r] Edin[r]

Plate IV.

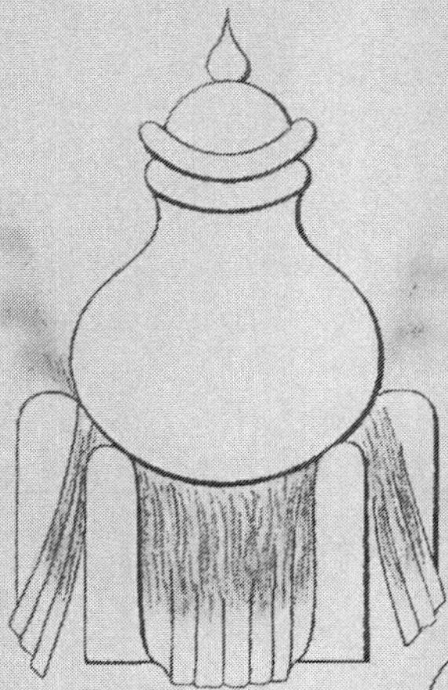

19. Koshti yantra No. II.

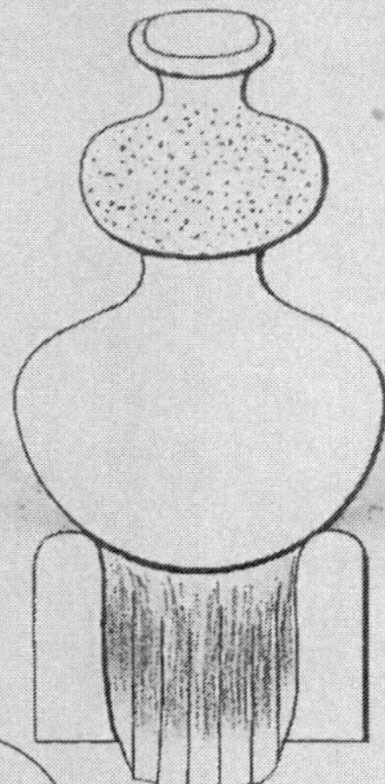

20. Lavana yantra.

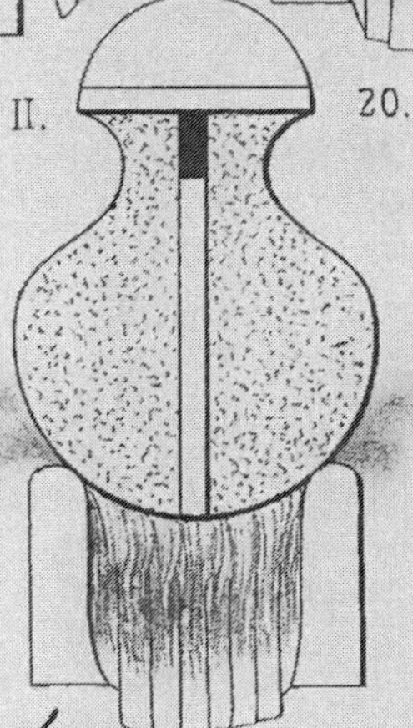

21. Nalika yantra.

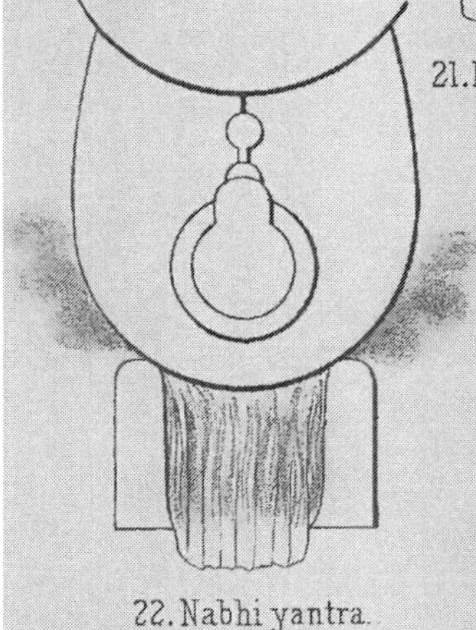

22. Nabhi yantra.

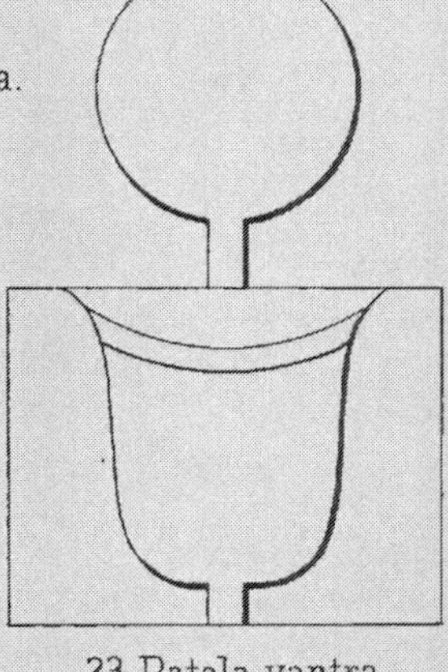

23. Patala yantra.

F. Huth, Lith[r] Edin[r]

Plate V.

24. Tiryakapatana yantra.

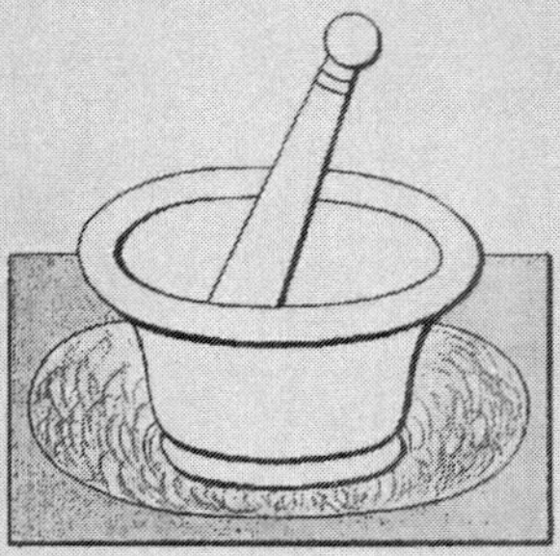

25. Taptakhalva yantra. No. I.

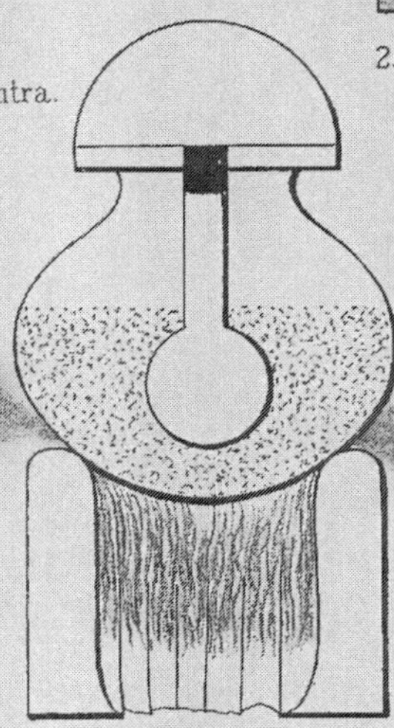

28. Valuka yantra.

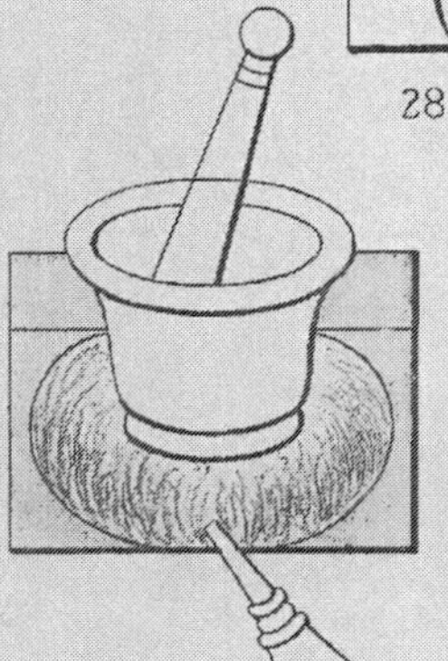

26. Taptakhalva yantra. No. II

27. Taptakhalva yantra. No. III.

P. Huth, Lith[r] Edin[r]

Plate VI.

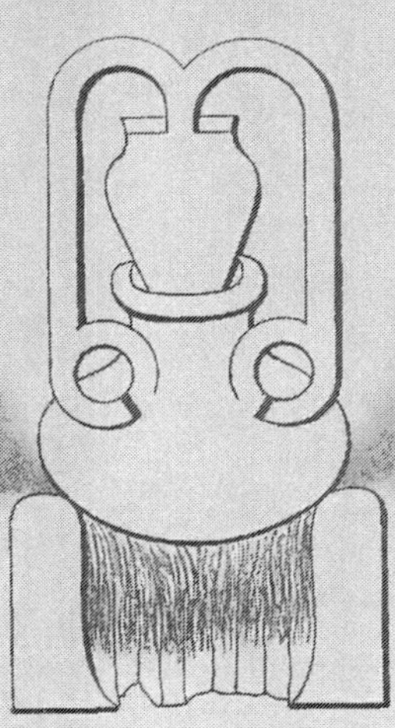

29. Valabhi yantra.

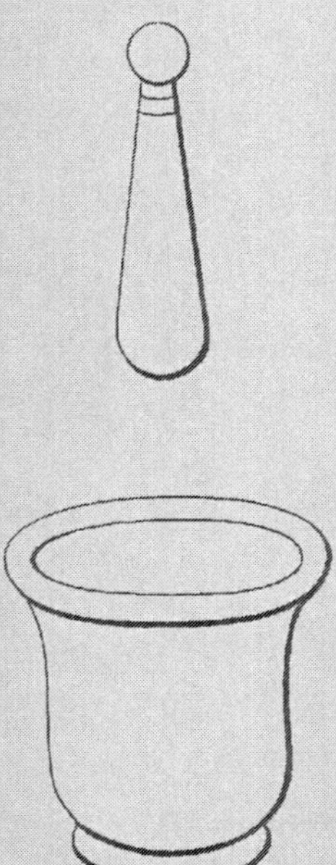

30. Vartulakhalva yantra.

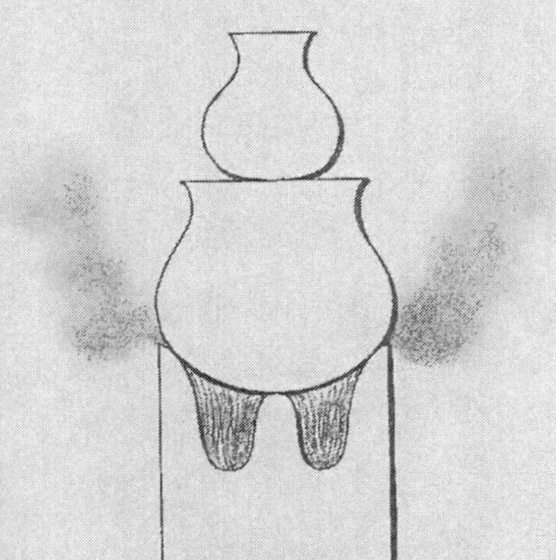

31. Vidyadhara yantra.

F. Huth, Lith. Edin.

moniac. Mould the lump in the form of a lozenge and let it dry. Apply to it asafœtida all round, and place it in an apparatus called Damaru (*see* Plate II.). Here, under the influence of fire, mercury will "fly up" and stick to the concave part of the covering vessel. That mercury is supposed to be perfectly pure. The next step in the preparation of its *Bhasma* is to take equal parts of dry ginger, black pepper, long pepper, carbonate of potash, barilla, "the five salts" (bay salt, table salt, Samber salt, black salt, and Bida salt), garlic, sal-ammoniac, bark of the horse-radish tree (Shigru), and mercury; powder together the dry ingredients first and then add the mercury, and thoroughly mix them together, triturating it for a whole day and night in Kanjika, or in lemon juice in Taptakhalva (*see* Plate V.). This process is called "Mukhakarana," which means literally the formation of a mouth; for mercury is then able to absorb any purified mineral sought to be blended with it. But it is not supposed to have acquired full strength until it is able to "imbibe" an equal, or double, treble, or even quadruple its quantity of prepared sulphur. Finally, triturate the purified mercury in betel-leaf juice, scoop out the interior of Karkoti root (*Dregia volubilis*) so as to form a cavity, fill it with the triturated mass, close the mouth, and drop the preparation in an earthen vessel luted with mud and cloth and place it over a fire. Mercury will then be fit to be reduced to *Bhasma*. There are other processes more or less intricate in which various Yantras or apparata are required.

Many treatises have been written on the wonderful power of Rasa or Mercury. The followers of a religious doctrine in India called Ruseshism consider Mercury as one of the manifestations of God. It is generally believed that the combination of mercury with other metals adds considerably to the intrinsic powers of these metals as useful remedies. Paracelsus of Hohenheim, who is known in Europe as "the Reformer of Medicine," in referring to the Yogis of India, says that "these are extremely long-lived, every man of them living to 150 or 200 years. They eat very little, rice and milk chiefly. And these people make use of a very strange beverage, a potion of sulphur and quicksilver mixed together, and this, they say, they drink twice every month. This, they say, gives them long life." The Yogis of India, as we have said, are supposed to have their own method of prolonging life to a wonderful extent by regulating their breathing. By a careful attention to the rules of conduct, diet, and ways of thought, as well as by the adoption of certain postures for restraining their inspirations and expirations, the Yogis are said to enjoy perfect health and happiness. They no doubt make use of certain medicinal drugs to stave off hunger and thirst without detriment to their health, and knowing its marvellous powers they now and then have occasion to use Parada (Mercury) during their austere practice. Sulphur being an indispensable ingredient in the preparation of the drug, Paracelsus is right in alluding to "a potion of sulphur and quicksilver." This "Luther Alter," as he is called, flourished in the sixteenth century, and had himself great faith in mercury, and his principal mixture was styled "Mercurius Vitæ." To mercury, when freed of all traces of lead, tin and other impurities, is ascribed the virtue of curing eighteen kinds of leprosy, eye diseases, fevers and impotency, and it is further credited with the power of prolonging life. As a therapeutic agent it is believed to be matchless.

Medicines prepared from the vegetable, animal,

and mineral products are exhibited in various forms. Some are used externally and some internally. These include :—

Anjana (eye-salve), Asava (tincture), Avaleha (electuary), Basti (enema), Bhasma ("alkaline ashes"), Bindu (drops), Chukra (vinegar), Churna (powder), Dhumapana (inhalation), Dhumra (fumigation), Drava (acid), Dravasveda (medicated bath), Fanta (infusion), Gandusha (gargle), Ghrita (ointment), Gutika (pill), Hima (watery extract), Kalka (paste), Kanjika (gruel), Kavalika (suppository), Kvatha (decoction), Lepa (plaster), Manjana (dentifrice), Modaka (bolus), Nasya (snuff), Paka (confection), Panaka (syrup), Peya (emulsion), Pindi (poultice), Plota (lotion), Satva (extract), Sechana (spray), Shrutambu (cold water in which a very hot piece of brick or iron is quenched), Shukta (ferment), Svarasa (expressed juice), Sveda (vapour), Taila (oil), Udvartana (liniment), Upanaha (fomentation), Vatika (lozenge) and Vartika (bougie).

The weights and measures employed in the preparation and administration of medicines have been referred to by the ancient Hindoo writers, and they are still used by the modern physicians, though the standard of weights prevailing in India varies in different parts. Charaka gives the following table for weights :—

| | |
|---|---|
| 6 Trasrenus . . . | = 1 Marichi. |
| 6 Marichis . . . . | = 1 Rajika. |
| 3 Rajikas . . . . | = 1 Sarsapa. |
| 8 Sarsapas . . . . | = 1 Yava. |
| 4 Yavas . . . . | = 1 Gunja. |
| 6 Gunjas . . . . | = 1 Mashaka. |
| 4 Mashakas . . . | = 1 Shana. |
| 2 Shanas . . . . | = 1 Kola. |
| 2 Kolas . . . . | = 1 Karsha (one tola).* |
| 2 Karshas . . . . | = 1 Ardhavapada. |
| 2 Ardhavapadas . . | = 1 Pala. |
| 2 Palas . . . . . | = 1 Prasruti. |
| 2 Prasrutis . . . | = 1 Anjali. |
| 2 Anjalis . . . . | = 1 Manika. |
| 2 Manikas . . . . | = 1 Prastha. |
| 4 Prasthas . . . . | = 1 Adhaka. |
| 4 Adhakas . . . . | = 1 Drona. |
| 2 Dronas . . . . | = 1 Surpa. |
| 2 Surpas . . . . | = 1 Droni. |
| 4 Dronis . . . . | = 1 Khari. |
| 100 Palas . . . . . | = 1 Tula (400 Tolas). |
| 2000 Palas . . . . . | = 1 Bhara. |

The weights used in Kalinga,—country along

* 180 Grains.

the Coromandel coast, north of Madras,—are the following :—

| | |
|---|---|
| 12 Sarsapas . . . . | = 1 Yava (barley-corn). |
| 2 Yavas . . . . | = 1 Gunja. |
| 3 Gunjas . . . . | = 1 Vala. |
| 8 Gunjas . . . . | = 1 Masha. |
| 4 Mashas . . . . | = 1 Shana. |
| 6 Mashas . . . . | = 1 Gadiana. |
| 10 Mashas . . . . | = 1 Karsha. |
| 4 Karshas . . . . | = 1 Pala. |
| 4 Palas . . . . | = 1 Kudava. |

The weights now in vogue in Gujarat and Kathiawar are :—

| | |
|---|---|
| 6 Chokhas . . . . | = 1 Rati. |
| 3 Ratis . . . . . | = 1 Vala. |
| 16 Valas . . . . . | = 1 Gadiana. |
| 2 Gadianas . . . | = 1 Tola (180 Grains). |
| 2 Tolas . . . . . | = 1 Adhola. |
| 2 Adholas . . . . | = 1 Navatanka. |
| 2 Navatankas . . . | = 1 Paseer. |
| 2 Paseers . . . . | = 1 Adhseer. |
| 2 Adhseers . . . | = 1 Seer. |
| 40 Seers . . . . . | = 1 Maund. |

Fluids were measured by a vessel of bamboo, wood, or iron, four fingers in diameter and as many deep, called Kudava.

# CHAPTER VIII.

## HINDOO WRITERS ON ÆTIOLOGY, DIAGNOSIS AND TREATMENT.

AFTER considering the Institutes of Medicine under the head of SUTRASTHANA, the early writers on Indian Medicine devoted their thought and attention to the investigation of the causes and symptoms of diseases, which they call NIDANA. Among the Hindoo writers, Madhava has investigated at length the causes and symptoms of the largest number of diseases in all their varieties. To this head Sushruta has devoted sixteen chapters treating of the classification, causation, and symptoms of diseases, such as diseases caused by wind, hæmorrhoids, urinary calculi, fistulæ, diseases of skin, urethral discharges, abdominal tumours, abortion and unnatural labours, abscesses, erysipelas, carbuncles, tumours, scrotal tumours, fractures, and dislocations, diseases of the male organs of generation, diseases of the mouth, and minor diseases. Sharngdhara enumerates eighty principal diseases caused by wind, forty by derangements of bile, twenty by abnormalities of phlegm, and ten by faulty conditions of blood. Sushruta traces all diseases to one or other of the following seven causes :—(*a*) corrupt semen virile or ovum of the father and mother respectively, causing leprosy, etc. ; (*b*) indulgence in forbidden food by the mother during pregnancy, or the non-fulfilment of any of her desires during that condition, causing blindness, etc., to the child ; (*c*) the derangement of humours in the body, causing fever, etc. ; (*d*) accidents, as fall, snake-bite, etc. ; (*e*) variations in the climate, causing cold, etc. ; (*f*) superhuman agencies ; (*g*) nature, as hunger, etc.

Harita reduces the number to three, and says that diseases are caused by Karma, or by the derangement of the humours, or by both. Karma is the unavoidable consequence of good or evil acts done in this or in a past existence. "Misery and happiness in this life are the inevitable results of our conduct in a past life, and our actions here will determine our happiness or misery in the life to come. When any creature dies, he is born again in some higher or lower state of existence, according to his merit or demerit." So there are certain diseases which are supposed to be the fruits of evil deeds done in a former state of existence. Harita declares that a murderer of a Brahman will suffer from anæmia, a cow-killer from leprosy, a regicide from consumption, and a murderer in general from diarrhœa. One committing adultery with his master's wife will suffer from gonorrhœa, and the violator of his preceptor's couch from retention of urine. A backbiter will suffer from asthma, a misleader from giddiness, a cheat from epilepsy, one who occasions or procures abortion from liver complaint, a drunkard from skin diseases, an incendiary from erysipelas, and one prying into another's secrets will lose the sight of one eye. Diseases caused by Karma may be cured by propitiatory rites, expiating ceremonies, and tranquillising efforts. If the rites do not cure the diseases, the patients have the assurance that they will at least check the further progress of the maladies in the life to come!

The treatment of diseases caused by humours forms part of CHIKITSA, or the therapeutic branch of Hindoo medical science. The Aryans have treated and prescribed remedies for different diseases, such as:—Diseases of the abdominal glands; abdominal tumours (of eight varieties); abortions (of six kinds); abscesses (of six kinds); anæmia; anorexia; apoplectic diseases; asthma; blood and bile affections; carbuncles (of nine varieties); cardiac diseases; cholera; colic (eight forms); convulsions; cough; cranial diseases; cystic affections; diabetes (eight types); diarrhœa (of seven varieties); dropsy; dysentery (of five kinds); dyspepsia; ear diseases; ectropium; enteric catarrh; entozoa; epilepsy (four varieties); erysipelas (of nine kinds); diseases of the eye, cornea, eye-balls, eye-lashes, eye-lids; diseases from excessive drinking; diseases caused by excessive thirst; fevers (twenty-five types); fistulæ (of eight sorts); fractures (eight forms); general debility; gonorrhœa (twenty varieties); hiccough; insanity (of four kinds); insensibility (of four kinds); jaundice; diseases of the lens; diseases of the male organs of generation caused by Shuka or water-leeches (of twenty-four varieties); mental debility; minor diseases (sixty sorts); diseases of the mouth (of seventy-four kinds); nose affections; paralysis (various forms); phlegm affections; piles (six forms); pustules and sores caused by urethral discharges (ten varieties); rheumatism; scrotal tumours (of seven kinds); skin diseases (eighteen forms); swellings (of nine sorts); sympathetic diseases; traumatic affections of the eye-ball; tumours; ulcers (fifteen varieties); unnatural labours; diseases of the urinary organs; urinary calculi (of four kinds); virile debility; vomiting; warts; diseases of wind; worm diseases (of twenty-one varieties); wounds (of eight kinds); and miscellancous diseases. Diseases of women form the subject of a separate chapter; infantile ailments and nursing are treated under the head of KAUMARBHRITYA; and symptoms and treatment of diseases supposed to be caused by superhuman powers are described under BHOOTAVIDYA.

The treatment of poisons and their antidotes come under the head of KALPA. Poisons are of two kinds, namely: STHAVARA, vegetable and mineral poisons; and JANGAMA, animal poisons. Datura, arsenic, and others are Sthavara poisons, and are cured by emetics, purgatives, errhines, collyria, and antiphlogistic treatment. Jangama poisons include venoms of such animals as insects, scorpions, spiders, lizards, serpents, mad dogs, foxes, jackals, wolves, bears, tigers, etc. Various antidotes are prescribed for different bites. Both kinds of poisons are used therapeutically by the Hindoos. Sometimes one poison is used as an antidote against another—*Vishasya visham aushadham*—as the dictum is —by administering a Sthavara poison to one suffering from the effects of a Jangama poison, and *vice versa.* A curious antidote is suggested by one writer, who says that the beating of a kettle-drum, besmeared with a preparation called "Ksharagad," before a person under the influence of poison, has the power of effecting a cure!

In diagnosing a disease, the Hindoo physicians have been guided from an early date by physical signs afforded by inspection, palpation,* percussion, auscultation, olfaction and degustation. Certain ancient writers take exception to the last, but others do not, and expect the physician to employ every one of his five senses, if necessary, in arriving at a correct conclusion regarding the seat and nature of a malady. The physician is required to note the patient's appearance, eye, tongue, skin, pulse, voice, urine, and fæces. The

* Palpation, percussion and auscultation are not altogether modern. They are referred to in the works of Charaka. Atreya, in his interesting dialogue with his favourite pupil Harita, speaks with even more precision on the subject. His directions are all of a piece with those in any of our modern works.

examination of the pulse is, however, considered the most important of all, as furnishing the best criterion of the phenomena and progress of disease, and it is the one usually depended upon by the native doctors. In order to know the precise character of the pulse, the radial artery at the wrist is usually chosen. In case of a male patient, his right pulse is generally felt, and in case of a female the left. In feeling the pulse the physician is to note its compressibility, frequency, regularity, size, and the different impressions it produces on the fingers. If it feels like the creeping of a serpent or a leech, wind is supposed to be predominant. If it be jumping like a frog, or similar to the flight of a crow or a sparrow, it indicates the predominance of bile. When it strikes the finger slowly and resembles the strutting of a peacock, it shows that the phlegm is in excess. The pulse that suggests the running of a partridge is called delirium pulse. An irregular pulse indicates *delirium tremens*, and a pulse which is almost imperceptible, depressed, irregular, and extremely languid, is a precursor of death. Pulsations in one suffering from fever or amorous passions are quick, and in a healthy man they are of a medium strength and perfectly regular. The capriciousness of the pulse produces other modifications very curiously described. It is interesting to note the similarity between this description of the pulse as found in the ancient Sanskrit treatises, and the doctrine of the pulse taught by the famous physician Galen, "who is the greatest and the best authority in Europe on the subject. For all subsequent writers have simply transferred his teaching on this subject bodily to their own works" (Dr Berdoe). Galen speaks of *pulsus myurus* (sharp-tailed pulse, so called as it sinks progressively and becomes smaller and smaller, like a mouse's tail); *pulsus formicans* (ant-like pulse, being scarcely perceptible, like the motion of an ant); *pulsus dorcadisans* (goat-leap pulse, as it seems to leap like a goat); *pulsus fluctuosus* (undulating), etc. This would suggest that Galen derived his knowledge on the subject from the works of Indian writers.

It has already been stated that certain kinds of diseases are believed to be caused by

> "Damned spirits all,
> That in cross-ways and floods have burial."

The demon theory of disease, a prevalent feature in almost every popular creed, has some influence on the Hindoo writers of medicine, according to whom the malignant spirit, if wittingly or unwittingly provoked, enters the body of the offender, annoys him in various ways, and afflicts him with certain kinds of diseases. Harita describes ten kinds of demons. Their names are: AINDRA, whose favourite resorts are monasteries, convents and shrines, manifests his mastery over a person by making him very emotional, or wild and furious; AGNEYA, who frequents cross-roads and burial-grounds, and under whose influence the patient looks intensely terrible and angry; NAIRUTI, who is found near ant-hills, and makes his victim either stand still or become violent; YAMA constantly seeks battle-fields and makes his victim excited; VARUNA haunts lakes and rivers, his victim looks like a dumb creature with watery eyes; MARUTA resides in whirlwinds, and the person possessed of him cries and shakes and feels otherwise excited; KUBERA makes his victim rash and conceited, exhibiting a passionate desire for ornaments; AISHANA has his abode in old temples, and under his influence the patient applies ashes to his body and moves about naked; GRAHAKA dwells in empty houses and dry wells, and one possessed of him cares not to eat or drink or listen to any one; PISHACHA is fond of dirty and unholy places, and his victim cries, sings, raves, and wanders naked like a madman.

Various kinds of medicinal and magical treatment are prescribed for demoniacal possessions. Sometimes amulets are tied round the neck of the patient. Here are specimens of four amulets generally used among the Hindoos for curing demoniacal and other diseases:—

| 39 | 66 | 96 | 73 |
|---|---|---|---|
| 11 | 44 | 32 | 34 |
| 23 | 93 | 23 | 94 |
| 46 | 26 | 4 | 64 |

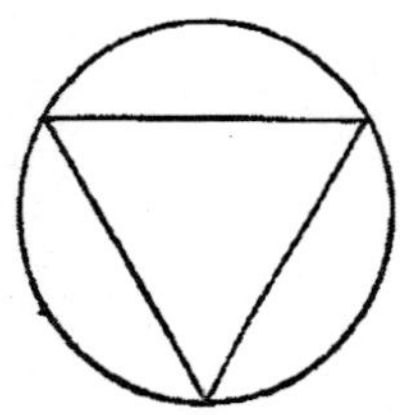

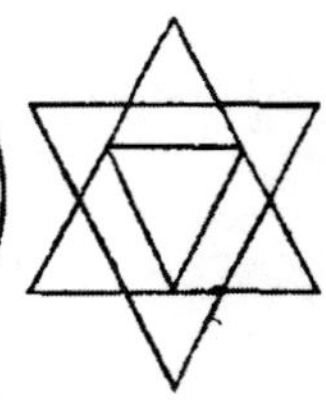

The first figure in which some mystic letters are written is used for exorcising a devil, the second is used for piles, the third for quieting a weeping child, and the fourth for fevers.

# CHAPTER IX.

## QUALITIES OF A PHYSICIAN AND HIS PROGNOSIS.

IF it is interesting to know the ideas of the early Indian writers on the theory and practice of medicine, and on other matters appertaining thereto, it will not be less interesting to describe their notions of what a good physician should be like. They have enumerated the qualities requisite in one desirous of practising as a doctor, and explained how he should behave both in private and public life in following his noble profession.

A physician is required to be always clean and tidy. For it is said that a physician who is dirtily and shabbily clad, conceited, foul-tongued, vulgar, and goes to a patient unasked, is not respected even though he be as clever as Dhanvantari. He should have his nails pared and his hair dressed, should have clean clothes, and carry a stick or an umbrella in his hand, wear shoes,* and have a gentlemanly bearing. He must be pure-minded, guileless, pious, friendly to all, and devoted to truth and duty. His chief duty is to treat his patient honestly, and without desire of any gain. To treat a patient conscientiously is supposed to bring "merit" (Punya) to the physician, who should not therefore sell his "Virtue" by treating a poor patient for the love of lucre. For the sake of his livelihood he will be justified in expecting an adequate fee from well-to-do people. He who is in a position to pay his doctor's fee but does not, though under his medical treatment, is styled "wicked," and is said to transfer all his "merit" to the physician. A religious sentiment appears to have been attached to the question of payment. For the Hindoos are enjoined not to approach or interview a king, a preceptor, and a physician "empty-handed," that is, without a gift or offering. It is therefore aptly said that a country is not without men, and men are not without diseases; so a physician's livelihood is always ensured.

* This recommendation will perhaps be thought superfluous by the European reader, to whom shoes are a necessary item of dress. In India, however, owing to climatic and other reasons, the covering for feet is not an indispensable article.

A practitioner knowing one hundred remedies for any one disease is called a Vaidya, one with a knowledge of two hundred remedies for any one disease is called a Bhishak, and to one who is acquainted with no less than three hundred remedies for each and every affection is applied the term Dhanvantari. The knowledge of diseases and the knowledge of the drugs are of equal importance to a physician. One without the other is like a vessel without a helmsman.

In the opinion of Sushruta, he who has merely learnt the principles of medicine, and received no practical instruction, loses his presence of mind when he sees a patient, just as a coward gets confused in a battle. On the other hand, he who through mere empiricism has obtained facility in practical work, but knows not the principles of medicine as taught in the books, deserves, not commendation from the learned, but punishment from the king. Both these are unaccomplished and unfit to become practitioners, just as a bird with a single wing is unable to fly.

Hindoo physicians go out to procure medicinal drugs on Sundays, Wednesdays, Thursdays, and Fridays of the light fortnight, and commence the preparation of mineral medicines on

Wednesdays, Thursdays, and Fridays. It is customary with some Vaidyas not to prescribe on Mondays, Wednesdays, and Saturdays—the first being particularly objectionable. Patients, on the other hand, avoid commencing treatment on Mondays, Tuesdays, and Saturdays, if they conveniently can. For purgatives or emetics Tuesdays, Thursdays, and Sundays, and for bloodletting Tuesdays and Sundays, are generally preferred.

After making the diagnosis, the physician forms an opinion as to the prognosis. Diseases are divided into three classes, namely, *Sadhya* (curable), *Asadhya* (incurable), and *Yapya* (controllable by remedies only). A patient suffering from a disease belonging to the last class remains well as long as he continues the use of medicine, but relapses as soon as the treatment is stopped, just as a tottering house collapses on the removal of the props. The physician is advised to refrain from treating a disease which is quite incurable.* The other two classes of diseases should be treated with all possible care and skill. In order to acquire success in his profession, the physician is expected to know both the theoretical and the practical sides of the science of medicine.

* Fortunately for the patient all are not of that opinion.

Omens are carefully watched by the Hindoo physicians before attending their patients. Favourable omens are such as the following: kettle-drum, tabour, conch, umbrella, cow with calf, virgin, woman with baby, two Brahmans, fish, horse, skylark, peacock, deer, mongoose, elephant, fruit, milk, flowers, dancing-girl, smokeless fire, flesh, spirituous liquor, sword, shield, dagger, washerman with dry-washed clothes, curds, cereals, banner, full water-pot, etc. The unlucky omens are fuel, hide, grass, smoky fire, snake, chaff, raw cotton, barren woman, oil, molasses, enemy, quarrelling people, one besmeared with ointment, scavenger, eunuch, butter-milk, mud, wet clothes, mendicant, ascetic, beggar, lunatic, one-eyed person, corpse, crow, jackal, empty water-pot, etc. These omens are observed with a view to enable the physician to prognosticate the favourable or unfavourable result of his attendance. They must be met with accidentally by the physician while on his way to the patient. But if the messenger who is sent to call the doctor sees on his way any of the omens enumerated above as good, it is bad for the patient; if he sees any of the bad ones it bodes good for the patient.

The messenger should preferably be of the same sex and caste as the patient, should be of good breeding, without any bodily deformity, clever, clean, well dressed, driving a horse or a bullock carriage, and holding fruits and white flowers in his hand. A widow or a mendicant is not considered a suitable messenger. When a physician is himself a qualified practitioner, well knowing the virtues and properties of drugs, his work becomes much easier if the messenger who comes for him is exact in his description of the disease, if the person attending the patient is careful in giving medicine at stated times, and if the patient is reasonable enough to follow the directions of the physician, and never to question the efficacy of the medicines prescribed.

Next to omens, the Hindoo physician seeks to derive some assistance from his knowledge of dream phenomena and astrology, to ascertain the probable result of his treatment. Everybody may be said to have experienced a dream, but few can say how the body in that state affects the mind, and how this affection produces the phenomena of dreams. Classical writers like Artemidorus, Macrobus, and Thomas Aquinas have in their works tried to solve the problem, and to establish the relation supposed to exist between the dreams and the events which they predict. The Indian writers, too, have endeavoured to throw some light on the question. This, however, is not the place to discuss their theory. Suffice it to say that the Indians have recognised dreams as the result of a state of life distinct both from the waking and the sleeping states, having at the same time a subtle connection with both. Assuming this to be the case, the Hindoo practitioners believe they can derive useful indications from the dreams of their patients. Dreams are, according to them, sometimes caused by fear, debility, and abnormal secretion of urine, wind, or bile. These are distinguished from those which are supposed to be prophetic and symbolic in their character. To ride a camel or a buffalo, to embrace a corpse or a mendicant, to see one's dead relatives, or find oneself besmeared with oil, to eat cooked food, or drink milk or oil, to see raw cotton, ashes, or bones, to discern a bare-headed black person riding a donkey and going in a southern direction, to find oneself decked with red flowers,—all such dreams are considered bad. A healthy man dreaming of these things will get ill, while a sick person will get worse. On the other hand, if one dreams of seeing a living king, friend, or a Brahman, sacred places, muddy water, mountains, rivers, elephants, horses, bees, leeches, or cows, or finds himself covered over with filth, blood, or flesh, or sees his own end approaching, he may hope to be prosperous if healthy, and to recover from sickness if ill. It is an unfavourable dream if a man suffering from fever associates with a dog; if a consumptive man sees an ape, a lunatic, or a demon; one suffering either from gonorrhœa, diarrhœa or dropsy sees water; or one subject to epileptic fits sees a dead body. If a leper drinks oil in a dream, or one with abdominal tumour dreams of eating vegetables, or one suffering from cold of eating buns, if one subject to asthma travels in dream, or an anæmic person dreams of eating yellow food, the results are equally unfavourable. The Indians believe in a deity called "Svapneshvari," or Goddess of Dreams, who is supposed to reveal certain events to her votaries in dreams. Remedies are prescribed for averting, as far as possible, the evil effects of dreams.

Astrology is considered to be a helpmeet to the medical science. The Aryans have from prehistoric times pinned their faith on the influence exercised on mankind by the nine planets, namely, the Sun,* Moon, Mars, Mercury, Jupiter, Venus, Saturn, *Rahu* (the Moon's ascending node), and *Ketu* (the Moon's descending node). They believe, in common with many other races, that these heavenly bodies rule the destinies of men and nations; and alleging that they possess a knowledge of their relative influence on the actions of each individual, they profess to be able to penetrate into his present, past, and future. Mr Proctor, in his well-known work, "Our Place among the Infinities," says "that of all the errors into which men have fallen in their desire to penetrate into futurity, Astrology is the most respectable, we may even say the most reasonable." It is also admitted that modern Astronomy owes a good deal of its early progress to Astrology. Kepler, in his preface to the Rudophine Tables, calls Astrology "a foolish daughter of a wise mother, to whose support and life the foolish daughter was indispensable." The admirers of the "daughter"

* According to the Hindoo belief the sun revolves round the earth, and not the earth round the sun. Hence it is that the sun is enumerated among the planets or the *Grahas* as they are called. The moon's ascending and descending nodes are considered obscure planets causing the eclipses of the sun and the moon.

have their own reasons to urge in her favour. It is beside our purpose to undertake to decide whether astrology is based on a scientific truth or is a relic of old superstition. Our present object is simply to record the fact that the Indian physicians are in the habit of consulting their patients' horoscopes when ordinary remedies fail to effect a cure. The malignant planets are appeased by various means. Mars, for instance, when he enters the house of the Moon, subjects the patient to blood-diseases. His evil influence is averted by reciting a certain sacred verse, by the gift of a red bullock to a learned Brahman,* and by an oblation of clarified butter in fire. Certain baths and wearing coral ornaments are also recommended under the circumstance. Different positions of the planets in the patient's horoscope are believed to have different effects, and the remedies vary accordingly. Predictions as regards the duration of a disease, or the possibility of its being cured or not, are now and then hazarded by certain Vaidyas from a consideration of the day of the fortnight or of the week on which the disease manifested itself. There are two opinions on the point, as

* Gifts are always made to Brahmans !

indicated in the following tables. The Roman numerals show the 1st, 2nd, 3rd, etc., days of the Hindoo fortnight, and the Arabic numerals, one below the other, show the number of days the disease will last (according to the two different authorities) if contracted on the day noted. The O indicates a fatal result.

| I. | II. | III. | IV. | V. | VI. | VII. | VIII. |
|---|---|---|---|---|---|---|---|
| 15 | 30 | 3 | 0 | 5 | 5 | 30 | 3 |
| 10 | 30 | 3 | 0 | 5 | 10 | 30 | 0 |
| **IX.** | **X.** | **XI.** | **XII.** | **XIII.** | **XIV.** | **Full Moon.** | **New Moon.** |
| 0 | 5 | 5 | 30 | 3 | 0 | 5 | 30 |
| 0 | 5 | 10 | 0 | 3 | 0 | 5 | 0 |

| Sun. | Mon. | Tues. | Wed. | Thur. | Fri. | Sat. |
|---|---|---|---|---|---|---|
| 30 | 45 | 0 | 8 | 5 | 7 | 30 |
| 0 | 45 | 0 | 8 | 6 | 7 | 0 |

Thus a disease beginning on the first day of the fortnight will last for fifteen days according to one writer, and ten days according to the other. If contracted on a Sunday it may last for thirty days or end fatally.

According to Sushruta, human life is either long, medium, or short. A long life may last for a hundred and twenty years, a medium life for seventy, and a short one for twenty-five years. One whose hands, feet, sides, back, nipples, teeth, shoulders, mouth, and forehead are large; whose arms, fingers, breath, and eyesight are long, brows and chest are broad; legs, genital organ, and neck short, voice and navel are deep; whose vigour is great, whose head protrudes backward, whose joints, veins, and arteries are buried in flesh, whose limbs are strongly built; who is cool and collected, free from disease, and has hair growing on the ears; whose body, intellect, and experience grow gradually,—such a man is expected to enjoy long life.

One expected to reach the medium age is said to have two or three wrinkles below his eyes; his legs and ears are fleshy and his nose is turned up.

Short fingers, a long sexual organ, a narrow back, conspicuous gums, and bewildered look, betoken a short life.

Sushruta also devotes a chapter to the description of what should go to make a symmetrical body.

Just as the outward form and bearing are supposed to enable the Indian physicians to say how long a person may be expected to live, there are certain signs and indications which it is believed enable them to conjecture when inevitable death will overtake him. Thus, if the breath flow through a man's right nostril continuously for one whole day, he will be no more after three years; if for two days, he will die in a year; and if it continuously flows from the same nostril for three days, he will not live beyond three months. If he breathes through the left nostril rapidly during the day and not at all during the night, he will die within four days. Again, he who breathes through the two nostrils simultaneously for ten days together will continue in life for three days only. If the right pulse is intermittent and the left nostril ceases to work, the patient is at the point of death. If his nose becomes bent, and if he is obliged to breathe through the mouth instead of through the nose, he will draw breath for thirty hours only. If one naturally dark suddenly becomes yellow he will die within two months. He whose teeth, lips, and tongue become dry, and eyes and nails black, and to whom yellow, green, and red appear black, will live for other six months only. He who sneezes at the time of the sexual orgasm and passes urine with the emission of semen, has only one year more of life. If fæces are voided simultaneously with urine, the patient will live for one year. If one's hands, feet, and chest dry immediately after coming out of the bath, death will take place in three months. A lean man suddenly becoming fat, or a fat man suddenly becoming lean, will die in six months. One unable to see the tip of his tongue will depart this life within twenty-four hours. If a miser becomes suddenly extravagant or charitable, it shows that he has only six months of life left. If half the body of a person remains warm and the other half cold, and if he has lost the power of hearing, death will overtake him in a week.

The duration of life is also ascertained by looking at the sun's reflection in a plate filled with water. If the patient finds the reflection entire and unbroken, he may be expected to recover soon; in case he finds it broken towards the south he will die in six months, if towards the west he will die after two months. If he finds it broken on the northern side his end will come in three months, if on the eastern side he will breathe his last in a month. If he sees a hole in the centre of the reflection he will expire before ten days are over, and the day he sees it surrounded by smoke will be his last.

# CHAPTER X.

## INDIAN SURGERY—ITS RISE AND FALL.

SHALYA or Surgery is, as noted in the earlier part of the work, one of the eight departments of AYURVEDA. In the work of Sushruta it occupies the first place. Medicine and Surgery, though parts of the same science, are treated as distinct branches. Charaka, Atreya, Hartia, Agnivesha, and others, are accepted as guides more in medicine than in surgery; while Dhanvantari, Sushruta, Aupadhenava, Aurabhra, Paushkalavata, and others, were rather surgeons than physicians, having written elaborate works on the art of healing by mechanical and instrumental means. In a case requiring surgical operations, the physician says to his patient, "*Atra Dhanvantarinam adhikaras kriyavidhau,*" meaning, "It is for the surgeon to take in hand this case." It is true the ancient surgery did not reach that perfection to which the modern science has attained. The successes of modern surgery are admitted on all hands to be prodigious, but that should not detract from the credit due to the ancients. The stock of surgical instruments and appliances used by the ancients was no doubt very small and meagre as compared with the armamentarium of a surgeon of the nineteenth century. The reason assigned for this fact is that the instruments they used were enough for their requirements, and that their acquaintance with the properties and virtues of drugs was so very great that most of the diseases and injuries now dealt with by the surgeon were then cured medicinally. An abscess, for instance, was either made to subside by certain kinds of plaster, or the swelling was assisted to mature by means of poultices, and when ripe was opened, not always with the knife, but by the application of a mixture of Danti, Chitrak, Eranda, and some other drugs. Cases of urinary calculi were treated with anti-lithics, and diuretics were administered so as to act as solvents for the stone, and thus the necessity of cutting was, if the patient so desired, obviated. It was only in rare cases, and for effecting a speedy recovery or affording immediate relief, that they had recourse to surgical operations. And yet their earliest works mention no less than one hundred and twenty-five surgical instruments for ophthalmic, obstetric, and other operations. They were experts in forming new ears and noses. This operation has been practised for ages in India, where cutting off the nose and ears was a common punishment, and "our modern surgeons have been able to borrow from them (Hindoos) the operation of rhinoplasty" (Weber). On this subject Dr Hirschberg of Berlin says:—"The whole plastic surgery in Europe had taken its new flight when these cunning devices of Indian workmen became known to us. The transplanting of sensible skin flaps is also an entirely Indian method." The same writer also gives credit to the Indians for discovering the art of cataract-couching, "which was entirely unknown to the Greeks, the Egyptians, or any other nation." The cataract operations are, it is said, performed by Indian practitioners with great success even to this day. The Hindoos were also experts in performing amputations and abdominal section. They could set fractures and dislocations in men and beasts, reduce hernia, cure piles and fistula-in-ano, and extract foreign bodies. Inoculation for small-pox seems to have been known to them from a very early age. Long before Edward Jenner was born, certain classes in India, especially cow-herds, shepherds, Charanas, and the like, had been in the habit of collecting and preserving the dry scabs of the pustules. A little of this they used to place on the forearm, and puncture the skin with a needle. In consequence of this inoculation, the classes are supposed to have enjoyed a certain amount of immunity from small-pox. Dr Huillet, late of Pondicherry, assures us that "Vaccination was known to a physician, Dhanvantari, who flourished before Hippocrates." The ancient Hindoos used to practise the dissection of the human body, and taught it to their disciples. They knew human anatomy and something of physiology. "The Hindoo philosophers," says Dr Wise, "undoubtedly deserve the credit of having, though opposed by strong prejudice, entertained sound and philosophical views respecting uses of the dead to the living; and were the first scientific and successful cultivators of the most important and essential of all the departments of medical knowledge—practical anatomy." It may as well be added that they were perfectly acquainted with the anatomy of the goat, sheep, horse, and other animals used in their sacrifices. Early warfare was conducted with such weapons as bow and arrow, sword, mace, etc. Thus in every war the services of bold and skilful surgeons were always in requisition for extracting arrows, amputating limbs, arresting hæmorrhage, and dressing wounds. Sushruta gives very minute directions to be observed in the performance of surgical operations, and describes the method of opening abscesses, treating inflammations, boils, tumours, ulcers and fistulæ, and of applying blisters, cautery, etc. The constant wars and internecine strifes afforded ample opportunities to the surgeons to distinguish themselves in their profession and acquire considerable dexterity in their work. A glance at the Vedic or the Epic period will bear testimony to this fact. The chirurgeons of yore are recorded to have performed incredible feats in surgical operations, just as modern surgery is able to do many things which ordinary folks will hardly believe to be possible. In its onward progress, modern surgery may yet be able to succeed in doing what the ancients claim to have performed. Sushruta classifies surgical operations into AHARYA, extracting solid bodies; BHEDYA, excising; CHHEDYA, incising; ESHYA, probing; LEKHYA, scarifying; SIVYA, suturing; VEDHYA, puncturing; and VISRAVANIYA, evacuating fluids.

The surgeon, before commencing an operation, is enjoined to equip himself with all the requisites, such as the instruments, salts, bandages, honey, oil, water, etc. He should have practical experience of his art, and should have seen many surgical operations performed by others. He should be intelligent, steady, skilful, and should execute his work with a light hand. He should have by his side steady and strong attendants to assist him. The patient should be allowed to take light food before any operation is performed upon him. Abdominal operations, however, and operations in the mouth, or about the anus, should be performed when the patient is fasting. The operation should be performed with the utmost care; and after it is over, a sesamum poultice should be applied io the wound, and a cloth bandage be tied round it. A certain incense should be kept burning ln the operation room. (This foreshadows the germ theory of the present day.) The surgeon should not leave his patient without offering a prayer to the Almighty for his speedy recovery. Particular attention is to be paid to the regimen of the patient. The wound must be dressed at regular intervals until it is all healed up. Should the wound cause intense pain, a cloth soaked in tepid ghee (clarified butter) mixed with liquorice may be applied to it.

As stated in the beginning of this chapter, the Indian surgery recognised 125 implements. These are grouped under two heads — YANTRAS (appliances) and SHASTRAS (instruments). The Yantras are 105, and are divided into six classes, viz. :—SVASTIKAS, pincers or forceps, twenty-four forms; SANDASHAS, tongs, of two sorts; TALAS, similar, of two kinds; NADIS, tubular instruments like catheters, etc., of twenty varieties; SHALAKAS, bougies, of thirty sorts; UPAYANTRAS, dressings, as cloth, twine, etc., twenty-six in number. These make a total of 104. The last, but not the least in importance, is the HAND, which is rightly considered to be the best and most indispensable implement in surgical operations. For specimens of some of the implements used in Indian surgery, refer to Plates VII.–VIII.

The SHASTRAS (instruments) are twenty in number, and are shown on Plates IX.–X.

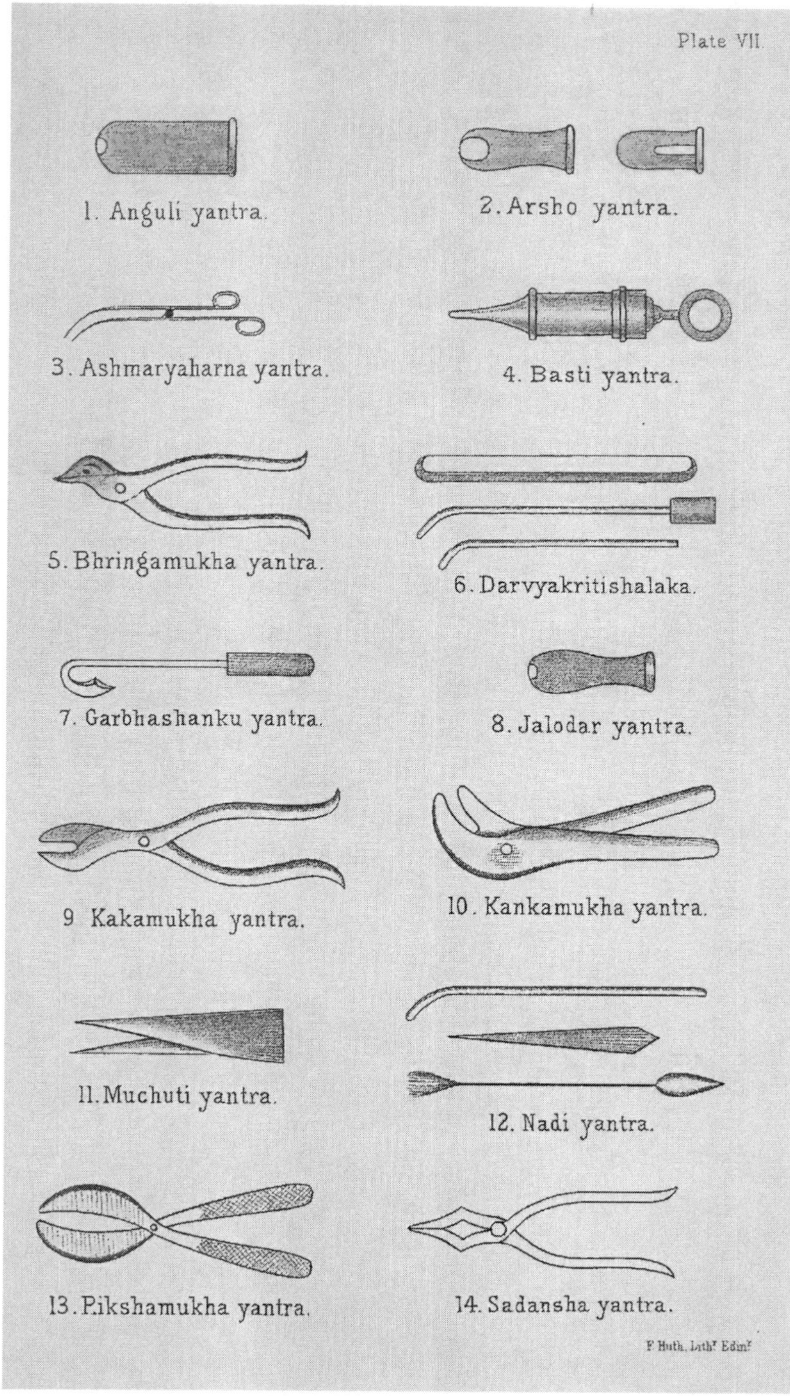
Plate VII.
1. Anguli yantra.
2. Arsho yantra.
3. Ashmaryaharna yantra.
4. Basti yantra.
5. Bhringamukha yantra.
6. Darvyakritishalaka.
7. Garbhashanku yantra.
8. Jalodar yantra.
9 Kakamukha yantra.
10. Kankamukha yantra.
11. Muchuti yantra.
12. Nadi yantra.
13. Rikshamukha yantra.
14. Sadansha yantra.
F Huth, Lith[r] Edin[r]

Plate VIII.

15. Shamipatra yantra.

16. Shalaka yantra.

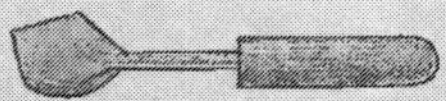

17. Sharapunka mukha.

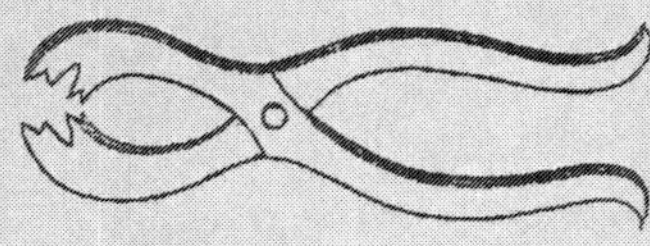

18. Sinhamukha yantra.

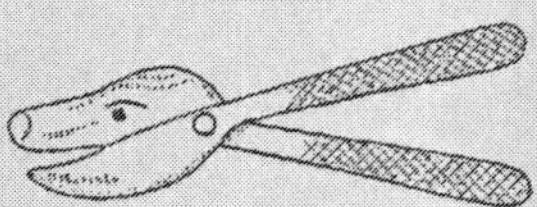

19. Shvanamukha yantra.

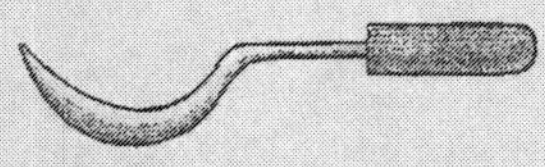

20. Shanku yantra.

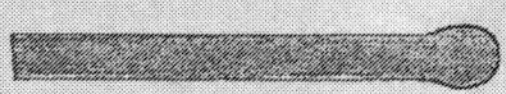

21. Snuhi yantra.

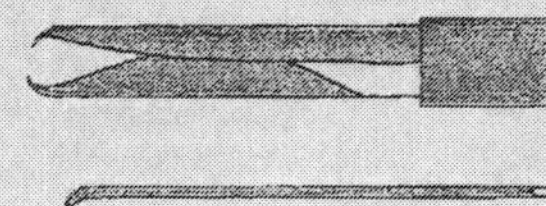

22. Tala yantra.

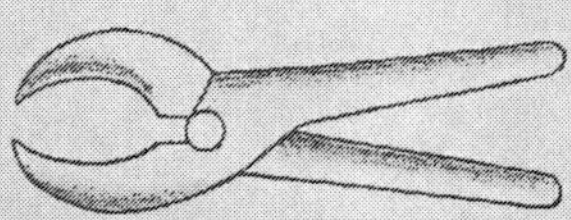

23. Tarakshumukha.

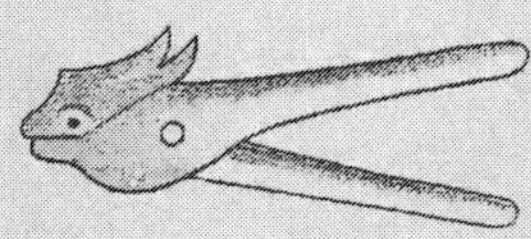

24. Vrikamukha yantra.

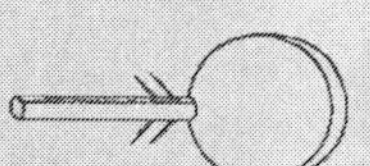

25. Vrinaprakshalana yantra.

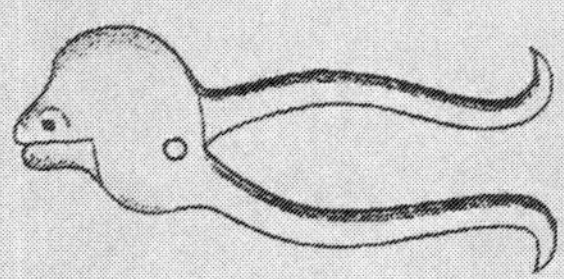

26. Vyaghramukha yantra.

27. Yugmashanku yantra.

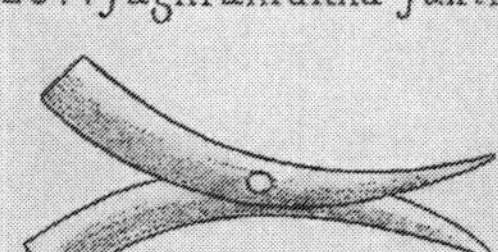

28 Yonyavekshana yantra.

F. Huth, Lith[r] Edin[r]

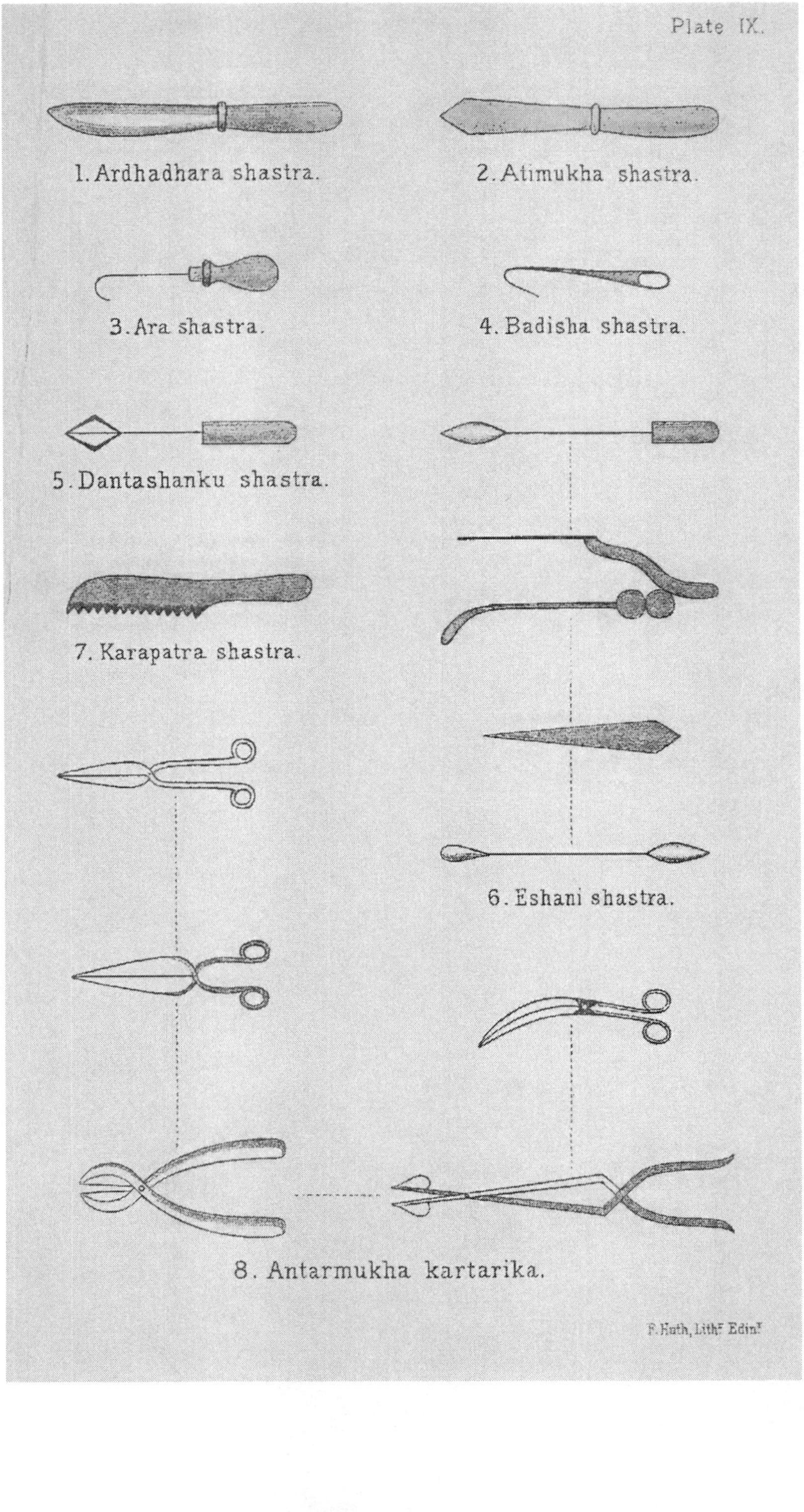
Plate IX.
1. Ardhadhara shastra.
2. Atimukha shastra.
3. Ara shastra.
4. Badisha shastra.
5. Dantashanku shastra.
7. Karapatra shastra.
6. Eshani shastra.
8. Antarmukha kartarika.
F. Huth, Lith.r Edin.r

Plate X.

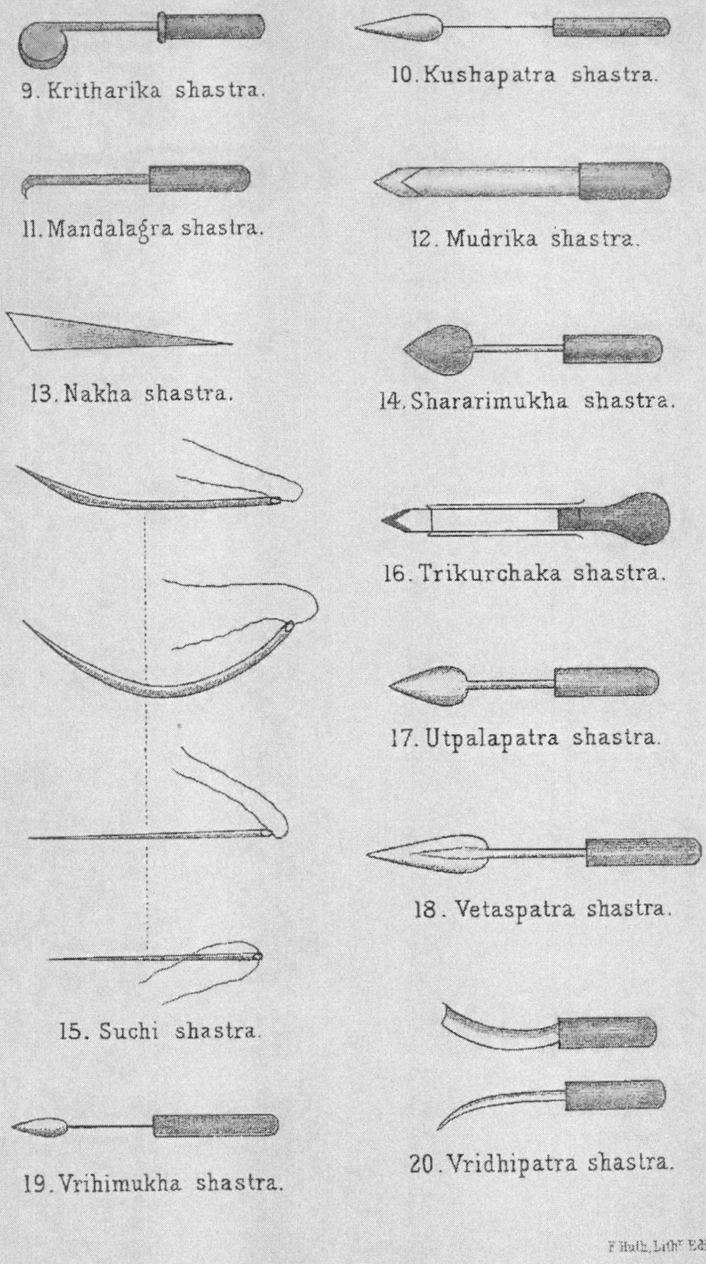


F Huth, Lith[r] Edin[r]

They are:—(1) ARDHADHARA, (2) ATIMUKHA, (3) ARA, (4) BADISHA, (5) DANTASHANKU, (6) ESHANI, (7) KARAPATRA, (8) KARTARIKA, (9) KRITHARIKA, (10) KUSHAPATRA, (11) MANDALAGRA, (12) MUDRIKA, (13) NAKHASHASTRA, (14) SHARARIMUKHA, (15) SUCHI, (16) TRIKURCHAKA, (17) UTPALAPATRAKA, (18) VRIDDHIPATRA, (19) VRIHIMUKHA, and (20) VETASPATRA.

The dimensions of these instruments are given in detail by old writers, who at the same time recommend that new implements and instruments should be introduced in accordance with the exigencies of the time and with the advice of experienced and competent surgeons. It is also enjoined that the instruments should be made of the best steel, for the manufacture of which India has been celebrated from the remotest times; they should be well shaped, with sharp, flawless edges, and should be kept in handsome, portable wooden boxes, with a separate compartment for each instrument. The surgical operations are performed on what are considered auspicious days. The patient is made to sit or stand with his face to the east, the surgeon before him with his face to the west. The surgeon should be cautious that no vital part, artery, vein, joint, or bone is carelessly injured in the course of the operation, and that the instrument does not go deeper than the requirements of the case actually demand. In serious surgical operations, and in diseases of a painful nature, the patient was made insensible by the administration of anæsthetics. In cases of children, or of patients having a dread of the knife, or where the proper instruments cannot be procured, bamboo, crystal, glass, Kurvinda (a kind of stone), leeches, fire, caustics, nail, Kareera (*Capparis aphylla*), Shefali (*Vitex Negundo*), hair and finger may be made use of. They are called ANUSHASTRAS or substitutes. Sharp pieces of bamboo bark or pointed crystal, glass, or Kurvinda may be employed as incisive instruments. The nail may be used in extracting a solid body, leeches in extracting blood, and hair, finger or vegetable sprout for probing. Caustics are used in opening abscesses, and fire (live charcoal) is applied to snake-bites and to wounds that are intensely painful. Thus there are three modes adopted by the Hindoos for treating surgical cases—by cutting instruments, by caustics, and by actual cautery. In the opinion of Sushruta, caustic is better than the knife, and cautery better than either.

In order to acquire dexterity in surgery, the preceptors made their pupils practise different operations on various substances. Incision, for instance, was practised on Pushpaphala (*Cucurbita maxima*), Alabu (*Langenaria vulgaris*), Kalinda (*Citrullus vulgaris*), Trapu (*Cucumis pubescens*), and other fruits; evacuating on a full Drita (a leather-bag for holding water), and on the urinary organs of dead animals; scarification on the fresh hides of animals on which the hair was allowed to remain; venesection was practised on the vessels of dead animals, and on the stalks of the water-lily; the art of probing and stuffing on bamboo, reed, cavities of wood and on dry Alabu; extraction of solid bodies on Panasa (*Artocarpus integrifolia*), Bilva (*Ægle Marmelos*), Bimbi (*Cephalandra indica*), and on the teeth of dead animals. "Removal of bad humours" (*scraping?*) was practised on wax spread on a board of Shalmali wood (*Bombax malabaricum*), and suturing on pieces of cloth, skin, or hide. Ligaturing and bandaging were practised on dummies; application of caustics and the actual cautery on pieces of flesh, and catheterisation on an unbaked earthen vessel filled with water.

The art of Surgery gradually declined in India

owing to a variety of causes, the chief among them being the aversion of the Brahmans, who had the monopoly of teaching the various sciences, to animal food and to the sacrificial offerings which were too common in the pre-Budhistic period. This aversion made them shrink from touching the carcasses necessary for anatomical demonstrations. They also shrank from coming in contact with blood, pus, and other matter, which cannot be avoided in performing surgical operations. Surgery being neglected by the priestly caste, passed into the hands of the lower classes, whose practice was purely empirical. Even these people, for want of encouragement, allowed it to decline, until, as Mr Elphinstone rightly remarks, bleeding was left to the barber, bone-setting to the herdsman, and the application of blisters to every man.

# CHAPTER XI.

## VICISSITUDES OF INDIAN MEDICINE AND SURGERY.

HINDOO MEDICINE was at the acme of its glory in the time of the Ramayana and the Mahabharata. To the court of every chief, great or small, was attached a physician, who was treated with great respect. There were Army Surgeons and Court Physicians. The work of the former was similar to that performed by the army surgeon of the present day. The Court physician used to wait upon the king every morning, and was the custodian of his health. Sushena was the name of the principal army surgeon of Rama in his war with Ravana, king of Lanka, and Valmiki makes mention of a particular Vaidya, who was Rama's personal physician. A similar practice is noticed in the time of the great war between the Pandavas and the Kauravas. The army surgeons were fully equipped with the necessary medical and surgical appliances (Bhishmaparva, Ch. 120). Duryodhana, the Chief of the Kurus, when pierced with arrows, was made by his surgeons to sit in a tub filled with medicated water, under which he was freed from the missiles lodged in his flesh (Mahabharat, Ch. 84). Both the conflicting armies had distinguished surgeons on their staff. Veterinary science seems to have been highly cultivated long before that period. Nala, a remote ancestor of the Pandavas, is described as a most accomplished horse-trainer, and as possessing a thorough knowledge of all matters relating to the horse. Nakula, one of the five Pandavas, was an expert in the veterinary science on which he has written several works, his "Ashva-chikitsa" being still extant. The science of treating elephants, bullocks and other domestic animals, was and is still known in India. Some are of opinion that Vagbhata, the celebrated author of "Ashtangahridaya," flourished in the time of the Mahabharata, and that he was the family physician of the Pandavas.

In the time of Buddha (B.C. 543), Indian medicine received the greatest support and stimulus, and surgery was allowed to languish. For Buddha and his followers would not permit the dissection of animals. They put a stop to animal sacrifice, in which a knowledge of anatomy was indispensable, and substituted models of dough. Buddha, however, established hospitals for men and beasts all over the country; and the institution of Pinjrapoles (Animal Hospitals), so peculiar to India, owes its origin to him.

The science continued to flourish down to the advent of the Greeks in India (B.C. 327). Arrian, the Greek historian, in describing the condition of India at the time of the invasion of Alexander the Great, refers to a curious fact, which reflects no small credit on the Hindoo physicians of the day. Alexander had in his train several proficient Greek physicians, but these had to confess their inability to deal with cases of snake-bite, very common in the Punjaub. Alexander was therefore obliged to consult the Indian Vaidyas, who successfully treated these cases. The Macedonian king was so struck with their skill that, according to Nearchus, he employed some good Vaidyas in his camp, and desired his followers to consult these Indian physicians in cases of snake-bite and other dangerous ailments. In face of the fact that the European toxicologists are still in search of a specific for snake-poison, the Indian physicians who lived some 2200 years ago might well be proud of their skill. It is very likely that on his homeward march Alexander, or Sikander as he is called in India, took with him a few professors of Hindoo medicine. This supposition receives some support from the early history of Greek medicine. There is a great similarity between the origins of the Greek and Indian medicine. Both the systems claim to be divinely inspired. The divine physicians Ashvins, the twin sons of the Sun, bear a close analogy to the divine twins Apollo and Artemis, who cured and alleviated the sufferings of mortals, and who derived their birth from Zeus, or the "God of Light." Hippocrates, the most celebrated physician of ancient Europe (B.C. 460), believed the art of medicine to be the production of the Divine Being; and it is curious to note that the Greeks, the Indians, and all the ancient nations of the world, have ascribed all kinds of knowledge, including that pertaining to the mysteries of life, disease, and death, to a superhuman agency. In the opinion of some writers, Hippocrates acquired his knowledge of medicine in India. The teaching of Pythagoras (B.C. 430), the founder of the Healing Art among the Greeks, is essentially Indian. He is said to have acquired his medical knowledge from the Egyptians, who, as will be shown further on, had borrowed their art from the Indians. Enfield, in his *History of Philosophy*, says that Pythagoras learnt his doctrine from Oriental philosophers, meaning the Hindoos. His philosophy bears such a striking resemblance to that of Buddha, that Mr Pocock, in his *India in Greece*, identifies him with "Buddhagurus" or Buddha. If he borrowed his philosophy from India, he may easily have borrowed the science of medicine from the same source. Plato and Hippocrates both believed in humoral pathology, and taught their pupils that the diseases in the body were caused by four humours,—blood, bile, phlegm, and water. The fact, however, that the three humours of the body are referred to in the *Rig Veda* (i. 34, 6), establishes the priority of the Indian system beyond all doubt. As for the Grecian physician Galen, who made himself famous at Rome in the second century of the Christian era, it has been said before that he adopted some of the fundamental principles of the Hindoo medical science in his works.

From these similarities one would be justified

in concluding that either the Aryans have copied their system of medicine from the Greeks, or the Greeks have derived theirs from the Indians. There is no internal or external evidence to support the first inference. For the Indians are a more ancient nation, and their medical books are older than any yet discovered on the surface of the earth. They are acknowledged on all hands to be thoroughly conservative, and as such have a natural repugnance to borrow. Sir William Hunter justly observes that Religion and Philosophy have been the great contributions of India to the world. As regards philosophy in general, Mr Colebrooke, in the *Transactions of the Royal Asiatic Society*, Vol. I., has reason to assert that "the Hindoos were teachers and not learners." All the important sciences have taken their birth in India. It does not stand to reason, therefore, to suppose that the science of medicine could have been borrowed from the Greeks, who themselves have lost all vestiges of that science, which is being practised at the present day all over India more or less in its original form. Professor Weber, who is never known to be partial to the Indians, asserts in his *History of Indian Literature* that "there is no ground whatever to suppose that Sushruta borrowed his system of medicine from the Greeks; on the contrary, there is much to tell against such an idea." The Indian books on medicine do not contain any technical terms which point to a foreign origin. Dr Hirschberg of Berlin, in a learned paper, adds, with regard to certain surgical operations, that "the Indians knew and practised ingenious operations, which always remained unknown to the Greeks, and which even we Europeans only learnt from them with surprise in the beginning of this century." Professor Diaz of the Königsberg University, clearly detects the principles of Indian medicine in the Greek system. Even those who talk eloquently of the antiquity of Greece withhold from her the credit of originality in regard to her medical science, and opine that the Greeks were indebted to Egypt for their knowledge of medicine.

The Aryans believe Egypt (Misra) to have been colonised by the Indians. Proofs are given in support of the belief, which it is beside our purpose to dilate upon here. Suffice it to say that the Tantrik deity *Nila-shikhandi* (black-crested), an incarnation of Rudra, is recorded to have first taught the Nilatantra (a mystical religious doctrine known to the Indians) in Egypt, the river Nile probably deriving its name from him. It is also stated that "in the reign of Vishvamitra, a certain king named Manu-vina, being excommunicated by Brahmans, emigrated with all his companions, passing through Arya (Iran or Persia), Baria (Arabia), and Misra (Egypt)." According to the Mahabharata, the four sons of Yayati, who were cursed by their father, migrated to the West, and became ancestors of some of the Mlechha tribes, and the name "Misra" (mixed) probably owes its origin to this circumstance. Sir William Jones, in the *Reports of the Royal Asiatic Society*, is led to believe that Egypt must have been in remote ages colonised by the Indian Aryans; and writers like Major Wilford consider the "Mishra-sthan" of the Purans to be no other than "Misra," the ancient name of Egypt. There is, on the other hand, no record of the Egyptians having ever migrated into India. Such circumstantial evidence has led some European writers—Louis Jacolliot among others—to affirm that if Egypt gave civilisation to Greece and the latter bequeathed it to Rome, Egypt herself received her laws, arts, and sciences from India.

There is nothing in the Egyptian medicine which is not in the Indian system, and there is much in the elaborate Indian system that is wanting in the medical science of Egypt.

It has been shown already that the Arab merchants took many medicinal drugs from India in the early part of the Christian era. It requires no great effort to prove that India has contributed greatly to the Arabic system of medicine. The Arabian physician Serapion (Ibn Serabi), in his well-known treatise upon Medicine, often quotes Charaka, who is named "Sharaka Indianus" in the Latin translation. Avicenna, better known by the name of Aflatoon in India—the name has become synonymous with a "learned man" among the Hindoos—flourished in the ninth century, and was the most celebrated physician of Bokhara. While describing the Indian preparation of Trifala (the three Myrobalans) in his work, he quotes the opinion of Charaka and other writers with great respect. Another Arabian physician, Rhazes (Al Rasi), who is said to have lived long before the two preceding physicians, in treating of the properties of dry ginger and other drugs, transcribes passages from the work of an Indian writer whom he calls "Sindhi-Chara." This Sindhi-Chara appears to be no other than the celebrated Vagbhata of Sindh, who was in his time known as a second Charaka or Chara—the syllable "*Ka*" making no difference, as in words like "bala" and "balaka," both meaning a child. The great works of Charaka and Sushruta were translated into Arabic, under the patronage of Khalif Almansur, in the seventh century. The Arabic version of Sushruta is known by the name of "Kelale-Shawshoor-al-Hindi." These translations, in their turn, were rendered into Latin. The Latin versions formed the basis of European medicine, which remained indebted to the eastern medicine down to the seventeenth century.

In the reign of King Vikrama (B.C. 57) Indian medicine was in the heyday of its glory. The ruler was a great patron of learning, and his court was made brilliant by the nine learned men, known as the "Nine Gems," a physician named Dhanvantari being one of them. It may be well to mention here that there have been several persons bearing the name of Dhanvantari, which is generally applied to an accomplished physician. The "gem" referred to as adorning Vikrama's court was the author of an elaborate work on Materia Medica, called *Nighantu.*

But perhaps there was no period in the history of Indian literature and science in which so liberal a patronage was given to learning in general, and to poetry and medicine in particular, as in the reign of King Bhoja of Dhar (A.C. 977). It was a golden age of Hindoo literature. The king was a learned man himself, and is the reputed author of a treatise on medicine and other works. Pandit Ballala, in his *Bhoja-prabandha*, or a collection of literary anecdotes relating to King Bhoja, describes an interesting surgical operation performed on the king, who was suffering from severe pain in the head. He tried all medicinal means, but to no purpose, and his condition became most critical, when two brother physicians happened to arrive in Dhar, who, after carefully considering the case, came to the conclusion that the patient would obtain no relief until surgically treated. They accordingly administered a drug called SAMMOHINI to render him insensible. When the patient was completely under the influence of the drug, they trephined his skull, removed from the brain the real cause of complaint, closed the opening, stitched the wound, and applied a healing balm. They are then related to have administered a restorative medicine called SANJI-

to the patient, who thereby regained consciousness, and experienced complete relief. This incident clearly shows that brain-surgery, which is considered one of the greatest achievements of modern science, was not unknown to the Indians. This is not a solitary instance. Jivaka, the personal physician of Buddha, is recorded to have practised cranial surgery with the greatest success. There are on record successful cases of abdominal section also. Thus it will be seen that the ancient Hindoos performed operations regarded as "triumphs of modern surgery." SAMMOHINI served the purpose of chloroform, but there is hardly a drug in the modern Pharmacopœia corresponding with SANJIVINI, which no doubt minimised the chances of "deaths under anæsthetics" that at present sometimes occur.

During the Mahomedan rule (A.C. 1001–1707), the Indian medicine began to show signs of decay. The reason is obvious. No art or science can flourish without the moral and material support of the government of the day. The Mahomedan conquerors brought with them their own Hakeems or doctors. The whole country was in an unsettled condition, not suitable for carrying on scientific investigations. The

Hakeems were an intelligent set of people. They unreservedly made use of some of the best and most effective Indian drugs, and incorporated them in their works. Among the important works written by the Hakeems may be mentioned *Al Fazl Advich*, by Noorudeen Mahomed Abdulla Shirazee, personal physician to the Emperor Shah Jahan (A.C. 1630). This work gives the names and properties of drugs sold in the Indian Bazaars; *Madan-us-shifa-i Sikandar*, by Beva bin Khas Khan (A.C. 1512), and *Tuhfat-ul-Muminin*, by Mahomed Momin, are compilations of the various Arabic and Sanskrit authorities on the science of medicine. Mahomed Akbar Arzani, court physician to Aurungzebe (A.C. 1658) in his *Karabadine Kaderi*, transcribes bodily many useful prescriptions from Sanskrit medical treatises. This shows that even in its decline the Hindoo medicine was able to command respect from its Mahomedan rival.

Indian medical science showed signs of revival during the time the Peshwas were in power (A.C. 1715–1818). The Peshwas came of high Brahman lineage, and they did all in their power to encourage indigenous learning and scholarship. All the learned men of the country were attracted to their court and liberally treated. Some of the recent works on Medicine, mostly compendiums of larger treatises on the subject, were written during this period.

The power of the Peshwas was overthrown by the English, and from the fall of the Marathas dates the decline of the native medical art, which lost all its material support. The English came with a pre-conceived notion that the Indian medicine was quackery, and the Hindoo works on the subject a repository of sheer nonsense. They established medical schools and colleges—an inestimable boon, no doubt—but looked upon the healing art of the land with supreme contempt. The Indians, on the other hand, with a natural dislike for everything foreign, supposed amputation and dressing of wounds to be the Alpha and Omega of the Western medical science.

# CHAPTER XII.

## CONCLUDING REMARKS.

MUCH of the misconception on both sides will disappear if the Hindoos care to remember that the English are one of the most progressive nations in the world; and the Englishmen bear in mind the words of Sir Monier Monier-Williams, who says:—"We are, in our Eastern Empire, not brought into contact with savage tribes who melt away before the superior force and intelligence of Europeans. Rather are we placed in the midst of great and ancient peoples who attained a high degree of civilisation when our forefathers were barbarians, and had a polished language, a cultivated literature, and abstruse system of philosophy, centuries before English existed even in name." If the question be approached with an open mind, without bias or prejudice, it will no doubt be found that the West, far more advanced though it be, may yet have something new to learn from the East. Those who have the advantage of being acquainted with both the systems are of opinion that, divested of all the exaggerations in which the Indians are prone to indulge, and of their tendency to consecrate all their sciences, and apotheosise their great men, the Hindoo system of medicine can, on the whole, bear comparison with the Western. There are many things in which both agree, and if in certain points they seem to differ, they often differ only to agree in the end. For instance, the wind-diseases of the Hindoos are mostly treated by the Western writers as diseases of the respiratory system; the bile-diseases generally correspond with the diseases of the circulatory system, and the disorders of the phlegm are analogous to the diseases of the alimentary system. The demoniacal diseases of the Hindoos are but other words for hysteria, epilepsy, dancing mania, and other disorders of the nervous system. It is also asserted by those who have had opportunities of learning and practising medicine, both on the Eastern and the Western principles, that Indian medical science has reached its highest standard of excellence in Materia Medica, Therapeutics and Hygiene, while the Western science is far more accurate and far superior in Chemistry, Anatomy, Physiology, and Surgery. The Indian science may well be proud of its symptomatology, diagnosis, and prognosis; and the Western science of its Pathology and Ætiology. The popular belief is, that in acute diseases European medicines are more effective than the Indian ones, but that in chronic cases the latter are more efficacious. In Legal Medicine the process of detecting poison by chemical analysis, resorted to by European toxicologists with great accuracy, is unknown to the Indians. In the preparation and administration of mineral drugs, the Hindoos claim to have a long experience. There is a striking resemblance between the two systems as regards the treatment of several diseases, such as diarrhœa, piles, asthma, consumption, paralysis, etc. It is but a truism to say that in some respects the Indian mode of treating certain diseases, peculiar to tropical climates, is more suitable and rational than any other. A close study of the science will convince an impartial reader that it contains germs of some of the modern discoveries in the healing art. A few of them, such as circulation of the blood, postural treatment, massage, and anæsthetics, have been referred to. A reference may also be made to the use of the magnet in therapeutics. Cures by animal magnetism were common in India long before they were recognised by Mesmer in Germany, and subsequently by John Elliotson in England. In the medical works of the Hindoos, doctors curing diseases by hypnotism are styled "Siddha" (endowed with supernatural power); those curing by means of mineral drugs "Daivi" (divine); those curing by vegetable preparations "Manushi" (human); and those by surgical operations "Rakshasi" (demoniacal). The names indicate the degree of estimation in which each class was held; and when Manu in his Ordinances directs his followers to "avoid the food of the doctor" (that is, to avoid eating with, or any food touched by a doctor), he evidently refers to the surgeons, and not to the other classes of physicians. The degenerate state to which Indian Surgery is now reduced is chiefly due to this popular prejudice.

The Indian writers have described the medicinal properties of waters of the principal rivers, lakes, wells, and mineral springs of the country, and their power to cure various diseases. This clearly shows that hydrotherapy was known in India long before it was dreamt of in Europe. It will thus appear that the Indian medicine does not deserve to be condemned off-hand. It has its faults, and its imperfections may be many, but it has also its good parts, few though they be. The aim and object of the two systems are the same. In the words of Charaka, "That is the true medicine, and that the true physician, that can cure and eradicate disease." Let the Western and the Eastern Schools of Medicine then join hands and reconcile themselves to each other wherever possible. Let them meet as friends, and not as foes or rivals. Under present circumstances, the East has much to learn from the West, but the West, too, may have something to acquire from the East, if it so chooses. If the Medical Science of India, in its palmy days, has directly or indirectly assisted the early growth of the Medical Science of Europe, it is but fair that the latter should show its gratitude by rendering all possible help to the former, old as it is, and almost dying for want of nourishment. The Indian Medicine deserves preservation and investigation. It is the business of all seekers after truth—be they

Europeans or Hindoos—to take up the question in the spirit of fairness and sympathy. The revival of such a spirit will, it is hoped, lead at no distant date to a juster appreciation of ARYAN MEDICAL SCIENCE.

# BIBLIOGRAPHY.

**Abhidhanaratnamala** describes several rare vegetable and mineral preparations.

**Abhrakakalpa**, by SHIVA, on the properties and medicinal uses of mica.

**Agnipurana**, a work compiled by VYASA, a section of which enumerates various drugs applicable to man and beast.

**Ajeernamritamanjari** by KASHIRAJ, on indigestion.

**Amritasagara**, with commentary and glossary by PRATAPSINHA on diseases in general.

**Anjananidana** by AGNIVESHA, with a commentary of Dattarama, a treatise on ophthalmology.

**Anupanatarangini** by PANDIT RAGHUNATHPRASADA on dietetics and regimen.

**Arogyachintamani** by PANDIT DAMODARA, on hygiene.

**Ashtangahridaya**, by VAGBHATA, in 120 chapters treats of anatomy, practice of medicine, surgery, ophthalmology, obstetrics, and hygiene.

**Ashtangasangraha**, a work on medicine by the same author. There are several commentaries on this work.

**Ashvavaidya**, by JAYADATTA SURI, on veterinary science.

**Atankadarpana** is a commentary on Madhavanidana by VACHASPATI, son of Pramoda, court physician of King Hamira.

**Atankatimirabhaskara**, by BALARAM of Benares. It treats of hygiene, nosology, astrology, diseases resulting from folly and vice, materia medica, and therapeutics.

**Aushadhanamavali**, by VAIDYA VIJAYASHANKAR, is a list of drugs alphabetically arranged.

**Ayurvedamahodadhi**, by SUSHENA, is an old treatise based on selections from the Ayurveda, and is highly valued by Hindoo physicians.

**Ayurvedaprakasha**, by SHRI MADHAVA of Benares in A.C. 1734, on the uses and preparations of vegetable and mineral drugs.

**Ayurvedavidnyana**, Hindoo system of medicine in two parts, compiled from Sanskrit treatises on medicine, surgery, chemistry, etc. by KAVIRAJ VINOD LAL SEN of Calcutta.

**Bhaishajyaratnavali**, by GOVINDADAS, on the theory and practice of physic.

**Bhavaprakasha**, by BHAVAMISHRA, in A.C. 1550. It is a summary of the practice of all the best Hindoo writers on medicine, and is the most popular work with Hindoo physicians all over India.

**Bhishaksarvasva**, a manual of medicine by an unknown author, treats of drugs applicable to a number of diseases.

**Bhojankutuhala**, by RAGHUNATHA SURI, on dietetics and regimen.

**Bopadevashataka**, by BOPADEVA, son of Keshava, 100 verses on practice of medicine.

**Brihatnighantaratnakara**, by DATTARAMA, on pharmacology

**Chakradatta** by CHAKRAPANI on drugs applicable to a number of diseases.

**Chamatkarachintamani**, by GOVINDARAJ, on marvellous remedies for various diseases.

**Charakasamhita**, by CHARAKA, a work of great antiquity on terms and definitions, the nature of diseases, remedies, on peculiar constitutions and temperaments, and diseases arising from them, materia medica, etc. The work is much sought after by the Hindoo practitioners, who refer to it as their best authority.

**Charyachandrodaya**, by DATTARAMA, on the laws of nature, their effects on human constitution, and on the means of preserving health.

**Charyapadmakara**, by RAGHUNATHPRASADA, on the same subject.

**Chikitsadhatusara**, by KASHINATH, on diseases in general, and also on minerals and metals.

**Chikitsakramakalpavalli**, by GOPALDAS, on practice of medicine.

**Chikitsasara**, by GOPALDASA, a short treatise on medicine, containing some useful formulæ.

**Chikitsasarasangraha**, by VANGASENA, on practice of physic.

**Dhatrimanjari**, by an unknown author, treats of pædiatrics.

**Dhatukalpa** forms a chapter of Rudrayamala, by SHIVA, on the therapeutic uses of metallic substances.

**Dhaturatnamala**, compiled from Ashvinikumarasamhita, on the preparations of metallic and mineral powders.

**Gandhakarasayana**, a useful treatise, devoted exclusively to the pharmaceutical preparations of sulphur.

**Hansarajanidana**, by HANSARAJA, on ætiology.

**Haritasamhita**, by HARITA, is an ancient work on the nature and treatment of diseases, pharmacy, properties of various kinds of food, water, climate, and diseases of women and children. The Hindoo practitioners hold the work in high veneration.

**Haritalakalpa** forms a chapter of Rudrayamala, by SHIVA, and describes the preparations and medicinal uses of yellow orpiment.

**Hastamalka**, by VAIDYA BAVABHAI, A.C. 1859, on the art of preparing metallic compounds.

**Hitopadesha**, by SHRIKANTHASHAMBHU, a diffuse treatise on diet and treatment of ordinary ailments.

**Jvaraparajaya**, by JAYARAVI, in A.C. 1794. It is a treatise exclusively on fevers.

**Kakachandeshvara**, on miraculous properties of mercury.

**Kumaratantra**, by RAVANA, king of Ceylon, on pædiatrics.

**Madanpalanighanta**, by MADANPAL, on materia medica.

**Madhavanidana**, by MADHAVACHARYA, in A.C. 1331, is an esteemed treatise on diagnosis.

**Mahapada**, by PALAKAPYA, an ancient physician. His work treats of elephants, their breeding and diseases.

**Mantratantraushadha**, compiled from Kamatantra, by SHIVA, on the uses of medicines prepared with the help of charms and incantations.

**Muktavali**, by CHAKRAPANI, on the nature and properties of medicinal drugs.

**Nadidarpana**, a small work on pulse by DATTARAMA.

**Nadidnyanatarangini**, a modern work on pulse by RAGHUNATHPRASAD PANDIT.

**Nadividnyana**, by DVARKANATHA BHATTACHARYA, an old treatise on pulse.

**Narayanavalokana**, by NARAYAN, on diseases caused by folly and vice, and treats of cures by means of charms, prayers, and incantations.

**Nayanasukha**, by a Jain priest, on diseases in general.

**Nidananjana**, by AGNIVESHA, an old work on diagnosis.

**Nighantaprakasha**, by JOSHI VAIDYA BAPU GANGADHAR, a useful dispensatory alphabetically arranged.

**Nighantaratnakara**, by VISHNU VASUDEVA GODBOLE, in A.C. 1867, an esteemed treatise on medicine and pharmacology.

**Pakavali**, by MADHAVA, treats of the preparations of different kinds of confections and alimentary substances.

**Paradakalpa** forms a chapter of Rudrayamala, by SHIVA, describing the preparations and medicinal uses of mercury.

**Paribhashavrittipradipa**, by GOVINDASEN, on the science of medicine.

**Pathyapathya**, by KAVI VISHVANATH, on dietetics and regimen.

**Pathyapathyanighanta**, by KAVI SRIMULLA, on dietetics.

**Prayogachintamani**, by MADHAVA, on pharmacy.

**Prayogasara** treats of diseases and their treatment.

**Ramavinoda**, by PADMARANGA, treats of the virtues of metallic substances and contains many recipes.

**Rasahridaya**, by Govind Bhikshu, court physician of king Madankirata, describes the process of preparing metallic and mineral compounds, with their uses as medicines.

**Rasamanjari**, by Shalinath, on treatment of diseases by mercurial preparations.

**Rasamrita**, by Pandit Vaidyakendra, in A.C. 1495, on pharmaceutical preparations of minerals and metals.

**Rasaparijata**, by Vaidya Shiromani, on pharmaceutical preparations in which mercury, arsenic, and certain metals are combined.

**Rasapradipa**, by Vishaldeva, in A.C. 1483, contains 500 stanzas, and describes the pharmaceutical preparations and uses of mercury.

**Rasaprakashasudha**, by Pandit Yashodhar, son of Padmanabha of Junagadh in A.C. 1550, on the pharmaceutical preparations and uses of metals and minerals.

**Rasarajashankara**, by Ramkrishna Mudgal, on the preparations of mineral drugs.

**Rasaratnakara**, by Ramachandra, on pharmaceutical preparations of metals and mercury.

**Rasarajasundara**, by Dattarama, on pharmaceutical preparations in which metals enter.

**Rasaratnasamuchchaya**, by Bhattacharya, son of Nripasinh Gupta, treats of pharmaceutical preparations of metals and minerals.

**Rasasanketakalika**, by Chamunda, a work of great antiquity, describing easy methods of preparing mineral drugs.

**Rasasara**, by Govindacharya, a short treatise on mineral remedies.

**Rasatarangamalika**, by JANARDANABHATTA, on the treatment of diseases by minerals.

**Rasayanaprakarana**, by SHRI MEDANUNGA SUREE, a Jain priest. He composed his work in Shrinagar in A.C. 1387, as appears from the manuscript. The work treats of the pharmaceutical preparations and uses of mineral and metallic substances.

**Rasendrachintamani**, by RAMCHANDRA BHATTA, on treatment of diseases by mineral drugs.

**Rasendramangala**, by NAGARJUNA, on the preparations and uses of mineral drugs.

**Rasendrasarasangraha**, with commentary of Hridayanath, by GOPALA BHATTA, on metals, gems, and pharmacy.

**Rasavatara**, by JAINACHARYA SIDHADIGAMBARA SHRI MANIKYADEVA, on the properties, purification, and oxidation of 'primary' and 'secondary' metals.

**Roganidana**, by DHANVANTARI, treats of diagnosis and describes different constitutions and temperaments.

**Sarvavijayeetantra**, by SHIVA, a work of great antiquity on the prolongation of life.

**Shalihotra**, by NAKULA, one of the five Pandavas, on the treatment of horses.

**Shandhachikitsa**, a treatise on the treatment of impotence.

**Sharngdharasamhita**, by SHARNGDHARA, on nosology and the practice of medicine.

**Sharngdharatika**, a commentary on Sharngdharasamhita by ADHAMULLA.

**Shatayoga** describes the preparations of various kinds of electuaries and decoctions, and enumerates one hundred modes of administering them.

**Sushrutasambhita**, by SUSHRUTA, with commentary of Dattarama one of the oldest medical works, is an abridgment of the Ayur Veda, and treats of anatomy, surgery, nosology, therapeutics, toxicology, and local ailments.

**Vaidyachintamani**, by DHANVANTARI, treats of nervous affections and derangements of the urinary system.

**Vaidyajeevena**, by LOLIMBRAJA, in A.C. 1633, a small treatise on practice of physic.

**Vaidyakakosha**, by DAUJI, with Hindi translation, is a dictionary of medicine.

**Vaidyakalpadruma**, by RAGHUNATHAPRASADA, a diffuse treatise on medicine.

**Vaidyamanautsava**, by BANSIDHARA, a treatise on medicine.

**Vaidyamrita**, by BHATTA MORESHVAR in A.C. 1627, on practice of medicine.

**Vaidyarahasya**, by VIDYAPATI, son of Bansidhara, in A.C. 1698, a compendium of the science of physic.

**Vaidyasarasangraha**, by RAGHUNATH SHASTRI DATYE and KRISHNA SHASTRI BHATAVEDEKAR, a general work on medicine.

**Vaidyavallabha**, by HASTI SURI, a Jain physician, in A.C. 1670, on simple treatment of diseases.

**Vangasena**, by VANGA PANDITA, a work of antiquity, treating of the preparations and uses of metallic substances and of diseases in general.

**Veerasinhavalokana**, by VEERSINHA, on nosology, and on diseases and their treatment by means of prayers, penance, charms, and charitable gifts.

**Yogachandrika**, by LAKSHAMANA, son of Pandit Datta, in A.C. 1633, on practice of medicine.

**Yogaratnakara**, by NAYANASHEKHARA, a Jain priest, in A.C. 1676, on the art of compounding.

**Yogaratnasamuchchaya**, by an ancient author, on diseases and their treatment by ordinary medicines as well as by means of incantations and charitable gifts.

**Yogashataka**, by PANDIT VARARUCHI, with a commentary by the Jain scholar Shridharasena on the nature and cure of certain diseases.

**Yogatarangini**, by TRIMULLA BHATTA, treats of materia medica, nosology, and pharmacology. It was composed in A.C. 1751.

# INDEX AND GLOSSARY.

PAGE

Abdomen, . . . . . 95, 98, 103
Abdominal glands and tumours, . . 151, 154, 168
section, proficiency of Hindoos in performing, 178, 198
Abhidhana Chudamani, a work on materia medica, . 120
Abhraka (mica), . . . . . 135
Abhyantara, internal, . . . . 38
Abortion, . . . . . 151, 154
*Abrus precatorius*, . . . . . 59
Abscess, . . . 151, 154, 177, 180, 184
*Acacia Catechu*, . . . . . 128
*arabica*, . . . . . 58
*farnesiana*, . . . . . 56
*Serissa*, . . . . . . 59
Accidental diseases, . . . . 55, 56, 57
Achara Madhava, a work on the usages of the Brahmans, 36
*Achyranthes aspera*, . . . . . 53, 59
Acids, . . . 14, 66, 82, 83, 137, 148
*Aconitum ferox*, . . . . . 110
*heterophyllum*, . . . . . 128
*Napellus*, . . . . . 109
*Acorus Calamus*, . . . 38, 52, 106, 110
Adhaka, a measure, . . . . . 149
Adhola, a measure, . . . . . 150
Adhseer, a measure, . . . . . 150
Adhyayas, sections, . . . . . 24
Ægineta, Paulus, . . . . . 124

PAGES

*Ægle Marmelos*, . . . . . 58, 185
Æsculapius, . . . . . . 32, 34
Ætiology, . . . . . . 203
Ætius, . . . . . . 124
Afghanistan, drugs imported from, . . . 126
After-birth, . . . . . . 49
Agada, antidote, . . . . . 26
Agantuka, accidental disease, . . . . 55
Agaru, *Aquilaria agallochum*, . . . 108
Agasti, a sage, . . . . . 30, 79
Agastya, a sage, . . . . . 68
*Agati grandiflora*, . . . . 108
Agent, therapeutic, . . . . 71, 113, 147
Agneya, a demon, . . . . . 159
Agni, god of fire, . . . . . 68
Agnidamani, *Artemisia sternutatoria*, . . 128
Agnivesha, disciple of Atreya, . 29, 34, 35, 110, 118, 176
Ague, . . . . . . 63
Ahankara, self-consciousness, . . . 41
Aharya, extracting solid bodies, . . . 180
Ahiphena, opium, . . . . 120, 126
Aindra, a malignant spirit, . . . . 159
Aishana, a demon, . . . . . 159
Ajakarna, *Shorea robusta*, . . . . 129
Ajamoda, *Ptychotis ajowan*, . . . 108, 129
Akakarabha, pellitory, . . . . 120
Akalkarabha, *Spilanthes oleracea*, . . . 106
Akasha, ether, . . . . . 42
Akhukarni, *Brachyramphus sonchifolius*, . . 110
Al Fazl Advich, a medical treatise, . . . 199
Al Rasi, . . . . . . 195
Alabu, *Langenaria vulgaris*, . . . . 185
Alexander the Great, . . . 15, 189, 190
*Alhagi maurorum*, . . . . . 128
Alimentary system, . . . . . 202

PAGES
*Allium Cepa*, . . . . . . 128
Alkaline ashes, . . . . . . 148
Almansur, a Khalif of Baghdad, . . . 196
Almond, . . . . . . 78, 120
Alochaka, one of the five kinds of Pitta, . . 89
Aloes, . . . . . 123, 141, 144
Aloe-smoking, . . . . . . 69
*Alstonia scholaris*, . . . . . 128
Alterative, . . . . . . 62, 117
Alu, *Prunus bocariensis*, . . . . 127
Alumina, silicate of, . . . . 135, 136
Ama, undigested food, . . . . . 96
Amblyopia, . . . . . . 102
Amla, sour property of, . . . . 102
Ammonium, chloride of, . . . . 136
*Amomum elettarum*, . . . . 108, 128
Amputation, . . . . 178, 180, 200
Amrita, *Cocculus cordifolius*, . . . . 108
fruit, . . . . . . 38
nectar, . . . . . . 34
Amukta, non-missive weapons, . . . 16
Amulet, . . . . . . 49, 160
*Amygdalus communis*, . . . . 111
*Amyris pentaphylla*, . . . . . 107
Anadi, eternal, . . . . . . 20
Anæmia, . . . 62, 94, 133, 142, 153, 154, 168
Anæsthetics, . . . 107, 184, 197, 198, 203
Analysis, chemical, . . . . 115, 202
Anannasa, pine-apple, . . . . 123
Anapatya, sterile, . . . . . 47
Anaphrodisiac, . . . . . 104
Anatomy, Hindoo, . . . . . 98
human, . . . . 179, 203
of animals, . . . . . 180
Anavalobhana, a religious rite, . . . 53

PAGES

Andaja, oviparous animals, . . . . 43
*Andropogon muricatus*, . . . . 67
*nardus*, . . . . 105, 128
*Schœnanthus*, . . . . 104, 128
Anecbolic, . . . . . . 107
Anemetic, . . . . . . 104
Angamardaprashamana, antispasmodic, . . 104
Animal heat, . . . . . 88
kingdom, drugs belonging to, . . 101, 148
Anjali, a measure, . . . . . 79, 149
Anjana, eye-salve, . . . . . 51, 148
Anjanee, mother of Hanumana, . . . 66
Anjira, fig, . . . . . 120
Annaprashana, a ceremony, . . . . 54
Anodyne, . . . . . . 109
Anointment, . . . . . . 51, 62
Anorexia, . . . . . . 154
Antalgide, . . . . . . 108
Antasthmatic, . . . . . 108
Anthelmintic, . . . . 106, 112
Antibromic, . . . . . . 107
Anticholeric, . . . . . . 107
Anticolic, . . . . . . 108
Antidote, . . . . 26, 65, 155, 156
Antifat, . . . . . . 106
Antihypnotic, . . . . . 106
Antilithics, . . . . . . 177
Antimony, . . . . . . 60
Antiphlegmagogue, . . . . . 105
Antiphlogistic treatment, . . . . 155
Antiphysical, . . . . . 107
Antipsoric, . . . . . . 105
Antipyretic, . . . . . . 105
Antipyrotic, . . . . . . 105
Antiscorbutic, . . . . . 106

PAGES
Antisialagogue, . . . . . . 106
Antisingultus, . . . . . . 105
Antispasmodic, . . . . . 104, 127
Antitoxic, . . . . . . . 110
Anulomana, cathartic, . . . . 104
Anupa, moist country, . . . . 56, 57
Anuretic, . . . . . . . 106
Anushastra, substitutes for surgical instruments, . 184
Ap, water, . . . . . . . 42
Apamarga, *Achyranthes aspera*, . . . 53, 59
Apana, a vital air, . . . . . 66, 87
Aphrodisiac, . . . . . . 69, 109, 139
Aphthæ, . . . . . . . 59
Appearance, to diagnose from, . . . 142
Appetite, . . . . . . . 65, 102
Appliances, . . . . . . 177, 182, 188
*Aquilaria agallochum*, . . . . 108
Aquinas, Thomas, . . . . . 166
Ara, a surgical instrument. *See* Plate IX., . . 183
Arabian travellers, works written by, . . 124
Arabic, Charaka and Sushruta translated into, . 34, 196
numerals, . . . . . . 171
Arabs, . . . . . . . 124, 143, 195
Arani, *Premna serratifolia*, . . . . 108
Ardha-chandra, a surgical instrument, . . 49
Ardhadhara, a surgical instrument. *See* Plate IX., . 183
Ardhavapada, a weight, . . . . 149
Ardhya, a kind of honey, . . . . 132
*Aristolochia bracteata*, . . . . 122
Arka, *Calotropis gigantea*, . . . 56, 58, 140
Army surgeons, . . . . . 29, 187, 188
Aromatics, use of, . . . . . 69
Arrian, the Greek historian, . . . . 189
Arsenic, . . . . . . 135, 140, 155
Arshogna, hœmostatic, . . . . . 104

PAGES
Artavotpadaka, emmenagogue, . . . . 104
*Artemisia sternutatoria*, . . . . 128
Artemidorus, . . . . . 166
Artemis, . . . . . . 190
Arteries, . . . . 46, 94, 157, 183
*Artocarpus integrifolia*, . . . 109, 185
*Arum Colocasia*, . . . . . 140
Arunadatta, Pandit, . . . . . 35
Arush-kara, *Semicarpus Anacardium*, . . 113
Arya, Persia, . . . . . . 194
Aryan, . . 11, 12, 15, 17, 18, 22, 153, 192, 193
circulation of the blood known to the early, . 93
humoral pathology of, . . . . 99
medicine, . . . . . 11, 205
physicians, . . . . 100, 134
Aryavarta, . . . . . . 12, 13
Asadhya, incurable, . . . . . 164
Asafœtida, . . . . . 107, 142, 145
Asana, *Bridelia tomentosa*, . . . . 50
Asanas, positions, . . . . . 75, 76
Asava, tincture, . . . . . 148
Asekya, kind of impotent person, . . . 47
Ashaya, hollow viscera, . . . . 96
Ashmarighna, lithonlytic, . . . . 104
Ashtadhyayi, a work on Sanskrit Grammar, . . 15
Ashtanga-hridaya, a medical work, . . . 34, 188
Ashtanga-sangraha, a medical work, . . . 35
Ashva-chikitsa, a work on veterinary science, . 188
Ashvagandha, *Physalis somnifera*, . . . 109
Ashvalayana, a sage, . . . . . 30
Ashvarudha, cavalry, . . . . . 16
Ashvatthama, a person of great age, . . . 58
Ashvini Kumars, sons of Soorya, . . . 26
Asita, a sage, . . . . . . 30
*Asparagus tomentosus*, . . . . 52

PAGES
Asthi, bone, . . . . . . 92, 95, 129
Asthma, . . 51, 61, 87, 130, 153, 154, 168, 203
Astika, one of the five "Happy Sleepers," . . 79
Astringent, . . . 66, 68, 82, 101, 103, 107, 114
Astrology, . . . . . . 169, 170
belief of the Hindoo physicians in, . 165, 169
Astronomical science, . . . . . 13, 169
Asuras, demons, . . . . . . 28
Atankatimirabhaskara, a medical work, . . 122
Atarusha, *Justicia Adhatoda*, . . . 128
Atasi, *Linum usitatissimum*, . . . . 129
Atharva Veda, . . . . . . 19, 23
Atimukha, a surgical instrument. *See* Plate IX., . 183
Ativisha, *Aconitum heterophyllum*, . . . 128
Atreya, an authority on medicine, . . 29, 159, 176
Atreya-Samhita, a medical work, . . . 29, 30
Audalaka, a kind of honey, . . . . . 132
Aupadhenva, a writer on surgery, . . . 176
Aurabhra, a writer on surgery, . . . 176
Aurungzebe, court physician of, . . . 199
Auscultation, . . . . . . 156
Avabhasini, a layer of the skin, . . . 97
Avalambana, a kind of kafa, . . . . . 90
Avaleha, electuary, . . . . . . 148
Avicenna, . . . . . . 34, 195
Avrishya, anaphrodisiac, . . . . . 104
Ayur Veda, "Science of Life," 23, 24, 26, 29, 31, 33, 34, 176
Ayurvedic literature, . . . . . . 37

B

Badakshani Naspasi, . . . . . . 38
Badarayana, a sage, . . . . . . 30
Badian, *Illicium anisatum*, . . . . 127
Badisha, a surgical instrument. *See* Plate IX., . 183
Bahirantara, exter-internal, . . . . 38

PAGES
Bahu-phena, *Sapindus emarginatus*, . . . 111
Bahya, external, . . . . . . 38
Bailly, . . . . . . . 14
Balakshaya, sterile, . . . . . . 47
Bali, . . . . . . . . 58
Balikhya, a sage, . . . . . . 30
Ballala, Pandit, . . . . . . 197
Balm, healing, . . . . . . 197
Bamboo, used in surgery, . . . . 184, 185
Banafsha, *Viola odorata*, . . . . . 127
Bandage, . . . . . . 25, 181, 185
Banga, tin, used as medicine, . . . 136
Banyan tree bark, . . . . . . 140
Baria, Arabia, . . . . . 126, 194
Barilla, . . . . . . . 145
Barley, . . . . . . . 82
Barren women, kinds of, . . . . 52, 165
*Bassia latifolia*, . . . . . . 116
Basti, enema, . . . . . . 148
*Batatas paniculata*, . . . . . . 109
Bath, . . . . . . 51, 62, 63, 81
Bavala, *Acacia arabica*, . . . . . 58
Bay salt, . . . . . . 144, 145
Beard, . . . . . . . 60
Bed-bug, used as medicine, . . . . 131
Beejpura, *Citrus acida*, . . . . . 107
*Berberis Lycium*, . . . . . . 128
Berdoe, Dr, . . . . . . . 158
Betel leaf, . . . . . 69, 78, 141, 145
    nut, . . . . . . 58, 69, 78
Beva bin Khaskhan, a physician, . . . 199
Bhagadeva, a Vedic personage, . . . 29
Bhagdugh, revenue collector, . . . 22
Bhagvat Gita, a sacred work, . . . 66, 68
Bhallataka, *Semicarpus Anacardium*, . 105, 112

PAGES
Bhanga, *Cannabis sativa*, . . . . . 110
Bhara, a weight, . . . . . . 149
Bharadvaja, a writer on medicine, . . . 30, 31
Bhargi, . . . . . . . 52
Bhasma, calx, 137, 138, 139, 140, 141, 142, 143, 146, 148
Bhava-mishra, author of a medical work, . 36, 37, 38, 120
Bhavaprakasha, a medical work, . . 30, 36, 94, 120
Bheda, a medical writer, . . . . . 29
Bhedana, purgative, . . . . . . 104
Bhedya, excising, . . . . . 180, 181
Bhela, disciple of Atreya, . . . . . 35
Bhishak, a physician, definition of, . . . 163
Bhishma-parva, a section of the Mahabharata, . 188
Bhoja, King of Dhar, . . . . . 197
Bhoja-prabandha, anecdotes about King Bhoja, . 197
Bhoomiti, geometry, . . . . . . 14
Bhoostrina, *Andropogon Schœnanthus*, . 104, 128
Bhootavidya, demonology, . . . . 25, 155
Bhrajaka, a kind of Pitta, . . . . 89, 97
Bhramara, a kind of honey, . . . . . 131
Bhringaraja, *Eclipta alba*, . . . . . 122
Bhudhara, an apparatus. *See* Plate I., . . 144
Bibheetaka, *Terminalia bellerica*, . . . . 105
Bile, . 83, 85, 88, 91, 102, 113, 115, 157, 167, 191, 202
of fish used medicinally, . . . 133
vessels for, . . . . . . 96, 97
Bilious temperament, . . . . . 90
Bilva, *Ægle Marmelos*, . . . . 58, 185
Bimbi, *Cephalandra indica*, . . . . 185
Bindu, drops, . . . . . . 148
Bite of mad animals, . . . . . 155
Bitter, . . . . 66, 68, 82, 101, 102, 114
Bitumen, . . . . . . . 135
Bleeding, . . . . . . . 186
Blindness, cause of, . . . . . . 152

PAGES
Blisters, . . . . . . 180, 186
Blood, . . 63, 69, 88, 89, 96, 168, 184, 186, 191
circulation of, . . . 92, 93, 94, 95, 203
formation of, . . . . 46, 89, 92
letting, . . . . 83, 131, 164
medicinal use of, . . . . 94
purification of, . . . . . 70
vessels, . . . . 92, 94, 97
Bodara, litharge, . . . . . 135
Body, foreign, . . . . . 25
symmetrical, . . . . . 172
*Boerhaavia diffusa*, . . . . 109, 140
Boils, . . . . . . 98, 180
Bolus, . . . . . . 148
*Bombax malabaricum*, . . . . 185
Bone, . . . . . 92, 96, 184
cuttlefish, . . . . . 129
formation of, . . . . . 92, 95
setting, . . . . . . 186
Bopadeva, a writer on medicine, . . . 39
Bopadeva shataka, a medical work, . . . 39
Borax, . . . . . . 35
*Boswellia thurifera*, . . . . . 123
Bougies, . . . . . 148, 182
Bowels, . . . . . 78, 86, 102
*Brachyramphus sonchifolius*, . . . . 110
Brahma, first writer on Hindoo medicine, . . 24, 26
a kind of diamond, . . . . 142
-vidya, highest philosophy, . . . 26
Brahmadandi, *Tricholapis glaberrima*, . . 121
Brahmans, . 12, 19, 54, 121, 153, 165, 186, 194, 199
Brahmi, *Hydrocotyle asiatica*, . . . 53
Breath, regulation of, . . . 117, 147
*Bridelia tomentosa*, . . . . . . 50
Brihadhara, a tree, . . . . . . 59

PAGES
Brihati, *Solanum indicum*, . . . . 105
Buddha, . . . . . 188, 191, 198
Buddhagurus, . . . . . 191
Buddhi, intelligence, . . . . 41
Burning sensation, . . . . 103, 113
*Butea frondosa*, . . . . . 128
Butter, . . . . 52, 58, 82, 130, 170
Butyracious substances, . . . . 66
*Byophytum sensitivum*, . . . . 122

C

Calcium, carbonate, . . . . . 136
Calmative, . . . . . . 108
*Calotropis gigantea*, . . . . . 56, 140
Camphor, paste of, . . . . . 63, 83
*Camphora officinarum*, . . . . 107
*Cannabis sativa*, . . . . . 110
*Capparis aphylla*, . . . . . 56, 184
*sepiaria*, . . . . . 106
Carbonates, . . . . 15, 135, 136, 145,
Carbuncle, . . . . . 151, 154
Cardamum, . . . . . . 69
*Cardiospermum Helicacabum*, . . . 104
Carminative, . . . . . 69, 105
Cartilages, . . . . . . 95
*Cassia Absus*, . . . . 113, 136
*fistula*, . . . . 109, 111
*lanceolata*, . . . . . 128
*Tora*, . . . . . . 112
Cassini, . . . . . . 14
Cataract-couching, . . . . . 178
Catarrh, . . . . . 112, 154
Catechu, . . . . . . 38, 69
Cathartic, . . . . . . 62, 104
Catheter, . . . . . . 182

PAGES
Catheterisation, . . . . . . 185
Cat's-eye, . . . . . . . 136
Caustics, . . . . . . 184, 185
Cautery, . . . . . . 180, 185
*Cephalandra indica*, . . . . . 185
Cerates, . . . . . . 132
Cerumen, . . . . . . 96
Chakradatta, . . . . . . 39
Chakshu, organ of sight, . . . . 41
Chala, motion, . . . . . . 94
Champaka, *Michelia Champaca*, . . . 59
Chandana, *Santalum album*, . . . 105, 129
Chandeshvar, father of Narahari, . . . 120
Chandrakanta, moonstone, . . . . 136
Chandramas, the Moon, . . . . . 29
Charaka, a writer on medicine, 29, 32, 33, 35, 37, 187, 195, 196
his definition of a true physician, . . 206
Charakasamhita, work by Charaka, . . . 33, 34
Charana, bard-caste, . . . . . 179
Charms, . . . . . . . 49, 160
Chassie, . . . . . . . 96
Chemistry, . . . . . . 14, 115, 203
Chhardinigrahana, anemetic, . . . . 104
Chhatra, a kind of honey, . . . . 132
Chhedana, laxative, . . . . . 105
Chhedya, incising, . . . . . 181, 185
Chikitsa, therapeutics, . . . . 153
Chikitsanjana, a medical work, . . . 39
Chikitsasangraha, a medical work, . . . . 39
Child, rearing of, . . . . . . 51
Childlessness, . . . . . . 33
Chitra-beeja, *Ricinus communis*, . . . 34, 129, 177
Chitraka, *Plumbago zeylanica*, . . . 116, 144, 177
Chlorides, . . . . . . . 15, 136
Chloroform, substitute for, . . . . . . 198

PAGES

Chlorosis, . . . . . . 65
Chokha, a weight, . . . . . 150
Cholagogue, . . . . . . 107
Cholera, . . . . . 141, 154
*Chrysanthemum Roxburghii*, . . . . 127
Chuda-karma, a religious rite, . . . 54
Chukra, vinegar, . . . . . 148
Chumbaka, loadstone, . . . . 135
Churna, powder, . . . . 58, 140, 148
Chyavana, a sage, . . . . 27, 28, 30
-avaleha, an electuary, . . . . 28
Chyle, . . . . . 46, 92, 93
*Cinamomum Cassia*, . . . . . 107
Circulatory system, . . . . . 201
*Cissampelos glabra*, . . . . . 49
*Citrullus vulgaris*, . . . . . 185
*Citrus acida*, . . . . . 107
Clarified butter, . . . . 52, 58, 182
Clay, used medicinally, . . 65, 134, 136, 139
*Clitoria Ternatea*, . . . . . 52, 110
Clove, . . . . . . 69
*Cocculus cordifolius*, . . . . . 108
*indica*. *See* Kakamachi, . . . . 144
Cocoa-tree and nuts, . . . . . 59, 124
Co-habitation, . . . . . 64, 73, 77
Cold, . . . . . 63, 91, 152, 168
Colebrooke, . . . . . . 14, 192
*Coleus aromatica*, . . . . . 112
Colic, . . . . . 87, 133, 134, 154
Collyria, . . . . . . 155
Conception, . . . . . 44, 53, 74, 96
Conch, . . . . . . 134
Conduct, rules of, . . . . . 77, 147
Conessi bark, . . . . . 59
Confection, . . . . . . 148

PAGES

Confinement, . . . . . 48, 50
Conjunctiva, application to, . . . . 60, 78
Constipation, . . . . 87, 127, 130
Consumption, . . . 61, 133, 153, 203
Contusion, . . . . . . 124
Convulsion, . . . . . . 154
Cooling food, . . . . 81, 83, 91, 113
medicine, . . . . 113, 114, 130
Copper, medicinal use of, . 133, 134, 135, 136, 139
Coral, . . . . . 133, 136, 170
*Coriandrum sativum*, . . . . 52, 109
Cornea, diseases of, . . . . . 154
Corrosive sublimate, . . . . . 136
*Cortex Ficus religiosa*, . . . . 106
Cosmetic, . . . . . . 109
Costive, . . . . 86, 87, 103, 116
*Costus speciosus*, . . . . . 104
Cough, . . . 51, 55, 113, 130, 133, 154
Countries, classes of, . . . . . 56
Cowry, . . . . . . 134
Cranial surgery, . . . . . 198
Cream, . . . . . . 130
*Cressa cretica*, . . . . . 121
*Crocus sativus*, . . . 63, 64, 78, 81, 108
*Croton polyandrum*, . . . . 116, 177
Crystal, medicinal use of, . . 65, 136, 184, 185
*Cucumis Colocynthis*, . . . . . 128
*pubescens*, . . . . . 185
*Cucurbita maxima*, . . . . . 185
*Cuminum Cyminum*, . . . . . 105
*Curcuma Zedoaria*, . . . . . 106
Curds, . . . 58, 73, 82, 83, 130, 165
Custard-apple, . . . . . 123

## D

PAGES

Dadhi, curds, . . . . . . 130
Dadhyanchi, a sage, . . . . . . 27
Dadima, *Punica granatum*, . . . . . 104
Dadru-ghna, *Cassia Tora*, . . . . . 112
Dahaprashamana, antipyrotic, . . . 105
Daivi, doctor curing diseases by mineral drugs, . 204
Dakshprajapati, a disciple of Brahma, . . 26
Dala, a kind of honey, . . . . . 132
Damaru-yantra, an apparatus. *See* Plate II., 144, 145
Dambha, escharotic, . . . . . . 105
Damodar, father of Sharngdhara, . . . 39
Dancing mania, . . . . . . 202
Danta, tooth, . . . . . . 130
Danti, *Croton polyandrum*, . . . 116, 177
Darshanas, Schools of Hindoo Philosophy, . . 16
Daru, *Pinus deodara*, . . . . . 50
Daruharidra, *Berberis Lycium*, . . . 128
Date fruit, . . . . . . . 38
  tree, . . . . . . . 59
*Datura Stramonium*, &c. *See* Dhattura, 106, 111, 128, 155
*Daucus Carota*, . . . . . . 105
Days, lucky and unlucky, . . 63, 64, 163, 164
Death, precursories of, . . . 173, 174, 190
  under anæsthetics, . . . . . 198
Debility, . . . . . . 154, 155, 167
Decoction, . . . . . . 144, 148
Deepaneeya, stomachic, . . . . 105
Deergha-fala, *Cassia fistula*, . . . . 111
Deglutition, . . . . . . . 88
Degustation, . . . . . . . 156
Delirium tremens, . . . . . . 157
Demoniacal possession, . . . 25, 160, 202
Demons, . . . . . . 158, 168
  amulet to exorcise, . . . 158, 159, 160

PAGES

Demulcent, . . . . . 109, 127, 131
Dentifrice, . . . . . . 148
Depilatory, . . . . . . 106
Devadali, *Luffa echinata*, . . . . 107
Devadatta, a vital air, . . . . 88
Devala, a sage, . . . . . 30
Devatas, gods, . . . . . 28
Devodas, King of Benares, . . . . 33, 34
Dhamanies, vessels, . . . . . 92, 97
Dhana, *Coriandrum sativum*, . . . 53, 109
Dhananjaya, a vital air, . . . . 88
Dhanur-ved, science of war, . . . . . 15
Dhanvantari, court physician of Vikrama, . . 196
  definition of, . . . . . . 163
  father of Indian medicine, 33, 34, 49, 119, 161, 176, 179, 196
Dhanvantari-nighantu, a work on materia medica, 119, 120, 121
Dhattura, *Datura Stramonium*, &c., . 106, 111, 128, 155
Dhatus, essential parts, . . . . 92, 94
  metals, . . . . . . 134
Dhaumya, a sage, . . . . . 30
Dhumapana, inhalation, . . . . 148
Dhumra, fumigation, . . . 133, 148
Diabetes, . . . . . . 25, 154
Diagnosis, . . . 36, 99, 142, 156, 164, 203
Diamond, bhasma of, . . . 142, 143
  medicinal use of, . . . 135, 142
Diaphoretic, . . . . . . 109
Diarrhœa, . 63, 87, 126, 130, 132, 153, 154, 168, 203
Diathesis, . . . . . . 86, 99
Diaz, Professor, . . . . . 193
Diet, . . . . . 68, 83, 147
Digestion, fire of, . . . . . 88, 114
  irregular, . . . . . . 86
  promotion of, . . . 69, 70, 102, 113, 116

PAGES

Digestion—*continued.*
retardation of, . . . . . . 67
Digestive canal, . . . . . 87
Dioscorides, . . . . . . 123
Discharges, . . . . . . 151
Discoloration, . . . . . 134
Discutient, . . . . . . 108
Disease, . . . . . 84, 85, 86, 164, 190
accidental, . . . . . 55, 56, 57
acute, . . . . . . . 203
of alimentary system, . . . . 202
apoplectic, . . . . . . 154
bile, . . 57, 82, 91, 96, 102, 127, 152, 154, 201
bladder, . . . . . . 89
blood, . 57, 62, 94, 103, 130, 133, 152, 154, 170
brain, . . . . . . . 142
causes and symptoms of, . . . 151
of chest, . . . . . . 61
of children, . . . . . . 155
chronic, . . . . . . . 203
of circulatory system, . . . . 202
cranial, . . . . . . . 154
curable, . . . . . . . 163
cure of, . . . . . . . 55
cystic, . . . . . . . 154
definition of, . . . . . . 155
demoniacal, . . . . 25, 158, 160, 202
duration of, . . . . . . 171
ear, . . . . . . 25, 63, 87, 155
from excessive drinking, . . . 154
eye, . 25, 63, 69, 130, 132, 133, 147, 154, 155
general, . . . . . . 25
heart, . . . . . . . 154
incurable, . . . . . . 164
internal, . . . . . . . 75

PAGES

Disease—*continued.*
kinds of, . . . . . 55
of lens, . . . . . . 154
of liver, . . . . . 140, 153
of male organs of generation, . . 151, 154
mental, . . . . . . 55, 154
minor, . . . . . . 152, 154
miscellaneous, . . . . . 154
of the mouth, . . . . 25, 151, 154
natural, . . . . . 55
nervous system, . . . . 131, 202
of the nose, . . . . . 25, 154
of parts above the clavicles, . . . 25
phlegmatic, . 56, 57, 65, 81, 91, 101, 152, 154, 202
physical, . . . . . 55
Portuguese, . . . . . 37
prevention of, . . . . . 55
rectal, . . . . . . 88, 103
respiratory system, . . . . 202
scalp, . . . . . . 61
scrofulous, . . . . . 130
seminal, . . . . . 88
skin, . . . 64, 117, 134, 151, 153, 155
splenic, . . . . . . 112
surgical, . . . . . 25
sympathetic, . . . . . 155
syphilitic, . . . . . 37, 127
theory of, . . . . . 84
traumatic, . . . . . 155
urinary, . . . . 88, 132, 147, 151
wind, . . . . . . 82, 202
of women, . . . . . 124, 155
Disinfectant, . . . . . . 134
Dislocation, . . . . . . 151, 178
Dissection, . . . . . . 179, 189

PAGES

Diuretic, . . . . . 106, 177
Doctor, English, . . . . . 127
kinds of, . . . . . 204
Mahomedan, . . . . 126, 199
respect for, . . . . . 73
*Dolichos sinensis*, . . . . . 109
Doshas, humours, . . . . . 85
Draksha, *Vitis vinifera*, . . . . 116
Drastic, . . . . . . 109
Drava, acid, . . . . . . 148
Dravasveda, medicated bath, . . . 148
Dreams, . . . . . 166, 167, 168
*Dregia volubilis*, . . . . . 146
Dressings, . . . . . . 182
Drita, a leather-bag for holding water, . . 185
Drona, and Droni, a measure, . . . 149
Drops, . . . . . . 148
Dropsy, . . . . 67, 130, 142, 154, 168
Dugdha, milk, . . . . . 130
Dung, . . . . . 134, 140, 142
Duryodhana, eldest of the Kuru princes, . . 188
Dvapara yuga, the third age of the world, . . 19, 35
Dvipa, continent, . . . . . 51
Dysentery, . . . . 55, 126, 132, 133, 154
Dyspepsia, . . . . . 63, 87, 154

E

Ear, cartilages of, . . . . . 95, 98
drops, . . . . . . 61
formation of new, . . . . 178
Earth, an element, . . . . . 41, 115, 208
Ecbolic, . . . . . . 105
*Echinops echinatus*, . . . . . 122
*Eclipta alba*, . . . . . 122
Ectropium, . . . . . . 154

PAGES
Egypt, medical system of, . . . 191, 194, 195
Ela, *Amomum elettarum*, . . . 108, 128
*Elæocarpus ganitrus*, . . . . . 121
Electuary, . . . . . 27, 52, 148
Elemental theory, . . . . . 115
Elivaka, aloes, . . . . . 69, 123
Elliotson, John, . . . . . 204
Elphinstone, . . . . . . 186
*Embelia ribes*, . . . . 53, 106, 112
Emerald, medicinal use of, . . . . 135
Emetic, . . . . 81, 107, 109, 155
Emmenagogue, . . . . . 104
Emollient, . . . . . . 130
Emulsion, . . . . . 132, 148
Enema, . . . . . . 148
Enfield, . . . . . . 191
Enteric catarrh, . . . . . 154
Entozoa, . . . . . . 154
Epilepsy, . . . . 141, 153, 154, 168, 202
Epispastic, . . . . . . 108
Epulotic, . . . . . . 107
Eranda. *See* Chitrabeeja, . . . 34, 129, 177
Errhines, . . . . . . 155
Eructation, . . . . . . 88
Eruptions, cutaneous, . . . . . 97
Erysipelas, . . . . 98, 151, 153, 154
Eschar, . . . . . . 113
Escharotic, . . . . . . 105
Eshya, probing, . . . . . 181
Essential parts, . . . . . 92, 96
Ether, one of the elements, . . . . 41, 115
*Euphorbia tirucalli*, . . . . . 108
European doctors, . . . . . 127
toxicologists, . . . . 189, 203
Evacuating, . . . . . . 181, 185

PAGES

*Evolvulus hirsutus*, . . . . . 107
Excising, . . . . . . 181
Excretion, . . . . . . 42
Exercise, physical, . . . 61, 62, 70, 81, 82, 83
Exhilarant, . . . . . . 69
Expectorant, . . . . . . 105
Expectoration, . . . . . 96
Expressed juice, . . . . . 148
Exsiccative, . . . . . . 105
Extract, . . . . . . 148
Extraction of foreign bodies, . . . . 180, 185
Extremities, . . . . . . 97
Eye, . . . . . 60, 95, 98, 154, 195
  averting the evil, . . . . 64
  cartilages of, . . . . . 95
  examination of, . . . . . 156
  lashes, . . . . . . 154
  pain in the, . . . . . 60
  salve, . . . . . 113, 148
  sight, improvement of, . . . . 102

F

Fainting, . . . . . . 103
Fall, . . . . . . . 55, 152
Fanta, infusion, . . . . . 148
Fat, . . . . 70, 92, 95, 102, 109, 132
  formation of, . . . . 46, 92, 95
Fatigue, . . . . . 61, 69, 70
Feather, . . . . . . 133
Fecundation, . . . . . . 75
Fenugreek, . . . . . . 120
Ferment, . . . . . . 148
Fetor, removal of, . . . . . 69
Fever, . 25, 55, 62, 103, 131, 133, 147, 152, 154, 168
  amulet for curing, . . . . . 160

PAGES

Fever—*continued.*
hectic, . . . . . . 142
intermittent, . . . . 117, 133
pulse in, . . . . . 157
*Ficus indica*, . . . . . 58
*religiosa*, . . . . . 59, 106
Fig, . . . . . . . 120
Figures, for exorcising devils, . . . 160
Finger, use of, in surgery, . . . . 184
Firanga-roga, Portuguese Disease, . . . 37
Fire, digestive, . . . 60, 66, 88, 95, 114
an element, . . . . 41, 42, 115
god of, . . . . . 68
used in surgery, . . . . . 184
Fissures, . . . . . . 134
Fistula-in-ano, . . . . . 178
Fistulæ, . . . . 129, 151, 154, 180
Fits, epileptic, . . . . . 168
Flaccidity, . . . . . . 102
Flatulence, . . . . . . 65, 134
Flesh, . . . . 70, 81, 92, 95, 165
formation of, . . . . . 31
Fly, swallowed to cause vomiting, . . . 131
Fœces, . . . . . 87, 96, 156, 174
Fœtus, . . . . . 45, 47, 87, 96
Fomentation, . . . . . 49, 148
Food, kinds and species of, . . . . 66, 67
undigested, . . . . . 96
Forceps, . . . . . . 182
Fractures, . . . . . 151, 154, 178
Frigorific, . . . . . . 109
Fumigation, . . . . . 133, 148

G

Gadiano, a weight, . . . . . 150
Gaduchi, *Tinospora cordifolia*, . . 50, 52, 129

PAGES
Gaetree, . . . . . . 56
Gajaputa, a furnace, . . . . . . 141
Gajarudha, warriors fighting on elephants, . . 16
Galactagogue, . . . . . 109
Galava, a sage, . . . . . 30
Galen, Claudius, . . . 114, 123, 158, 191
Gambier, a plant, . . . . . 120
Gamut, . . . . . . 15
Ganas, groups of medicinal plants, . . . 103
Gandha, smell, . . . . . 41
Gandhaka, sulphur, . . . 135, 139, 140, 146, 147
Gandusha, gargle, . . . . . 148
Ganita, mathematics, . . . . 14
Gaozaban, *Onosma bracteatum*, . . . 127
Garala, venom, . . . . . 130
Garbhadhana, a religious rite, . . . 53
Garbhasravi, ecbolic, . . . . . 105
a woman who conceives but miscarries, . . 47
Garga, a sage, . . . . . 30
Gargle, . . . . . . 148
Garlic, . . . . . 144, 145
Garutmat, emerald, . . . . . 135
Gautama, a sage, . . . . . 30
Gavedhu, a kind of corn, . . . . 106
Gems, . . . . 34, 135, 136, 196
Genætology, . . . . . . 78
Germ, . . . . . . 46
theory, . . . . . . 181
Gestation, period of, . . . . . 47
Ghanta-pushpa. *See* Dhattura, . 106, 111, 128, 155
Ghee, clarified butter, . . 53, 58, 130, 182
Ghrita, clarified butter, . . . . 130
ointment, . . . . . . 148
Ghritakumari, Indian aloes, . . . 141, 144
Giddiness, . . . . . 113, 153

PAGES

Ginger, . . . 50, 52, 58, 65, 145, 195
Glands, larger, . . . . . . 97
mammary, . . . . . . 96
Glass, medical use of, . . . 65, 136, 184
*Gloriosa superba*, . . . . . . 49
*Glycyrrhiza glabra*, . . . . . . 109
Gobhila, a sage, . . . . . . 30
Go-karni, *Clitoria Ternatea*, . . . . 110
Gokshura, *Tribulus terrestris*, . . . 104
Gold, . . . . 47, 64, 134, 138, 139
Goldstücker, . . . . . . . 20
Gomeda, onyx, . . . . . . 135
Gonorrhœa, . . . 127, 132, 141, 153, 154, 270
Gopichandana, silicate of alumina, . . . 136
Gorochana, a pigment, . . . . . 58
Gout, . . . . . . . 140
Graha, planet, . . . . . . 169
Grahaka, a demon, . . . . . . 159
Grahana, organ of smell, . . . . . 42
Grahi, carminative and exsiccative, . . . 105
Greek, 11, 20, 32, 34, 70, 99, 124, 178, 189, 190, 192, 193
medicine, . . . . . . 126, 190
Greeshma, the hot season, . . . . 80, 81
Grinjana, *Daucus Carota*, . . . . 105
Gruel, . . 49, 138, 140, 141, 142, 144, 145, 148
Guava, . . . . . . . 123
Guggula, *Amyris pentaphylla*, . . . . 107
Gul-e-daudi, *Chrysanthemum Roxburghii*, . . 127
Guna, quality, . . . . 67, 68, 101, 103
Gunja, *Abrus precatorius*, . . . . . 59
a weight, . . . . . 149, 150
Gutika, pill, . . . . . 137, 148
*Gyrardinia heterophylla*, . . . . . 121

## H

PAGES
Hæmorrhage, . . . . . . 79, 180
Hæmorrhoids, . . 79, 133, 154, 160, 178, 203
Hæmostatic, . . . . . . . 104
Hair, . . 60, 63, 96, 102, 103, 131, 184, 185
formation and growth of, . . 46, 61, 130
Hakeem, Mahomedan physician, . . 126, 199
Hand, chief implement in surgery, . . . 182
Hanumana, a deity, . . . . . . 58, 66
Haridra, *Curcuma Zedoaria*, . . . . . 106
Haridruma, gambier, . . . . . 120
Harishchandra, a physician, . . . . 63
Harita, a writer on medicine, 29, 49, 52, 94, 152, 156, 176
Haritaki, *Terminalia Chebula*, . . 53, 104, 114
Harita-Samhita, a medical work, . . 29, 30, 94
Haritshyama, turquoise, . . . . . 136
Harvey, William, . . . . . . 93, 94
Hasti, a writer on medicine, . . . . . 30
Head, . . . . . . 60, 72, 73, 90
ache, . . . . . . . 55, 79
Healing, . . . . . . . 41, 103
Health, . . . . . . . 73, 147
Heart, . . . . . 46, 85, 89, 90, 94
Heat, internal, . . . 65, 70, 88, 95, 114
Heating, . . . . 68, 69, 80, 81, 89, 113
*Hedychium spicatum*, . . . . . 105
Heera, diamond, . . . . . . 135
*Helekteres isora*, . . . . . 108
Hemanta, the frosty season, . . . . 80, 83
Henna, . . . . . . . 123
Hepatic functions, . . . . . . 65
Hepler, . . . . . . . 34
Herbs, . . . . . . 80, 103, 124
Hernia, . . . . . . . 178
*Heydysarum gangeticum*, . . . . . 108

PAGES

Hiccough, . . . . . 87, 133, 154
Hikkanigraha, antisingultus, . . . . 105
Hima, watery extract, . . . . . 148
Hindoo, . 11, 12, 13, 14, 15, 17, 19, 25, 73, 83, 84
  circulation of the blood, theory of the, . . 92
  humoral theory, . . . . . 99
  knowledge of contagion, . . . 72
  law, . . . . . . 16, 36
  medical works, . . . . 192, 200, 203
  pharmacopœa, . . . . . 131
  philosophy, . . . 16, 17, 201, 224
  postural treatment, . . . . 75
  practice of shampooing, . . . . 89
Hindoos, 11, 13, 15, 114, 115, 153, 156, 179, 202, 205
  first to use mercury internally, . . . 143
  practised as physicians in Baghdad, . . 124
  traditional and unwritten lore of, . . 117
  used poisons as medicines, . . . 155
Hingu, *Narthex Asafœtida*, . . . 107, 142, 145
Hingula, red sulphide of mercury, . . 135, 141
Hintala, a tree, . . . . . . 59
Hiouen Thseang, . . . . . . 84
*Hipian orientali*, . . . . . . 122
Hippocrates, . . . 99, 115, 123, 179, 190, 191
Hiranyakeshi, a sage, . . . . . 77
Hirschberg, Dr, . . . . 178, 193
Homeopathists, . . . . . . 113
Honey, . . . . 38, 52, 131, 132, 181
Horn, . . . . . . . 134
Horse radish tree, . . . . . . 145
Hospitals, . . . . . . 128, 189
Hot diet, . . . . . . . 68, 83
Huillet, Dr, . . . . . . . 179
Humidity, . . . . . . . 82
Humoral pathology, . . . . . 191, 152

PAGES
Humours, . . 57, 65, 86, 90, 91, 99, 101, 185, 191
derangement of, . . 85, 100, 108, 152, 153
equilibrium of, . . . . . 86
Hunger, . . . . . 55, 99, 147
Hunter, Sir W. W., . . . . 15, 34, 62, 192
Hydragogue, . . . . . 107
*Hydrocotyle asiatica*, . . . . 52
Hydrotherapy, . . . . . 204
Hygiene, principles of, . . . 55, 56, 84, 203
Hypersecretion of mucus, . . . . 60
Hypnotic, . . . . . . 106
Hypnotism, . . . . . . 204
Hypochondriasis, . . . . . 127
Hypogastrium, . . . . . 48, 87
Hysteria, . . . . . 131, 202

I

Ibn Serabi. *See* Serapion, . . . . 34, 195
Ikshu, *Saccharum officinarum*, . . 105, 108
*Illicium anisatum*, . . . . . 127
Immortality, . . . . . . 34, 42
Implements, number of, . . . 178, 182
Impotency, . . . . . 45, 133, 147
Impregnation, . . . . . 44, 75
Incense, in sick room, . . . . 181
Incision, . . . . . 181, 185
India, 15, 16, 34, 70, 71, 103, 149, 188, 191, 195, 199
materia medica of, . . . 117, 118
medical science of, 11, 75, 85, 117, 124, 168, 189, 191, 192, 195, 198, 199, 200, 203, 205
Indian brain surgery, . . . . 198
drugs, . . . . . 123, 195, 199
flora, . . . . . . 125
nuts, . . . . . . 124
philosophy, . . . 16, 17, 201, 224

PAGES

Indians, . . . . . . 12, 166
Indigestion, . . . . . . 62
Indra, king of the gods, . . 26, 27, 28, 29, 31
Indravaruni, *Cucumis Colocynthis*, . . . 128
Indrayava, *Wrightia antidysenterica*, . 104, 129
Inebrient, . . . . . . 91, 106
Inflammation, . . . . 79, 134, 180
Inflation of a corpse, . . . . . 88
Infusion . . . . . . 148
Inhalation, . . . . . . 148
Inoculation, . . . . . . 179
Insanity, . . . . . 55, 133, 154
Insect, bite of, . . . . . 155
Insecticide, . . . . . . 131
Insemination, . . . . . 74
Insensibility, . . . . . 154, 184
Inspection, . . . . . . 156
Inspiration, . . . . . . 147
Instruments, . . 25, 177, 178, 181, 182, 183, 184
Intelligence, . . . . 41, 52, 65, 72, 89, 102
Intercourse, sexual, . . . . . 72, 73
Intermittent fever, . . . . 117, 133
Intestine, . . . . . . 46, 88
Intoxicants, . . . . . . 91, 106
Intumescence, . . . . . . 65
Ionian system, . . . . . . 126
*Ipomoea Terpethum*, . . . . . 107
Iron, . . . 134, 135, 136, 141, 142, 147
Iron-legs, . . . . . . . 28
Irritation, . . . . . . . 60, 64
Irshyaka, a kind of impotent person, . . 47
Isphgul, *Plantago Ispaghula*, . . . 122
Itch, . . . . . . . 112

## J

PAGES
Jacolliot, Louis, . . . . . . 194
Jaggery, . . . . . . . 81
Jala, cobweb, . . . . . . 131
Jalas, plexuses, . . . . . . 97
Jaluka, leeches, . . . . . . 131
Jamadagni, a sage, . . . . . . 30
Jangala, dry country, . . . . 56, 57
subacetate of copper, . . . . 136
Jangama, animal poison, . . . 155, 156
Jarayu, viviparous animals, . . . . 43
Jasmine, . . . . . . . 59
Jatakarma, a ceremony at birth, . . . . 53
Jatamansi, *Nardostachys jatamansi*, . . 108, 123
Jatifala, *Myristica moschata*, . . . . 106
Jatukarna, a writer on medicine, . . . . 29
Jaundice, . . . . 65, 131, 133, 141, 154
Jeeraka, *Cuminum cyminum*, . . . . 105
Jeevata, living creatures, used as medicines, . . . 131
Jenner, Dr Edward, . . . . . 179
Jhulla-pushpa, *Byophytum sensitivum*, . . . 122
Jihva, organ of taste, . . . . . 41
Jijbal, *Xanthoxylum rhetsa*, . . . . 58
Jinjhira, *Triumfetta angulata*, . . . . 121
Jivaka, a physician, . . . . . . 198
Joints, . . . . 70, 90, 96, 127, 132, 184
Jones, Sir William, . . . . . . 194
Jotishmati, *Cardiospermum Helicacabum*, . . . 104
Juice, . 62, 80, 88, 89, 139, 140, 141, 144, 145, 148
Jujube, . . . . . . . . 57, 59
Jungle, forest, . . . . . 118, 125
Jupiter, . . . . . . . . 169
*Justicia Adhatoda*, . . . . . . 128
Jvarantaka, *Ophelia Chiretta*, . . . 112, 129

PAGE
Jvarhara, antipyretic, . . . . . 105
Jyotisha, astronomy, . . . . . . 13

K

Kacha, glass, . . . . . 65, 136, 184
Kadamba, *Nauclea Cadamba*, . . . 59
Kadara, *Acacia Catechu*, . . . . 128
Kafa, phlegm, . . . . . 85, 86, 89
Kafa-hara, antiphlegmagogue, . . . 105
Kafakara, phlegmagogue, . . . . 105
Kakajangha, *Capparis sepiaria*, . . . 106
Kakamachi, *Cocculus indica*, . . . . 144
Kaka-vandhya, a form of sterility, . . . 47
Kala-Madhava, a work on astronomy, . . 36
Kalas, receptacles, . . . . . . 96
Kalinda, *Citrullus vulgaris*, . . . . 185
Kalinga, a country along the Coromandel coast, . 149
Kali-yuga, the fourth age of the world, . . 19, 35
Kalka, paste, . . . . . . 148
Kalpa, toxicology, . . . . . . 155
Kamala, *Nelumbium speciosum*, . . . 107
Kama-sutra, a work by Vatsayana, . . . 74
Kamya, a sage, . . . . . . 30
Kandaras, large vessels in body, . . . 97
Kandughna, antipsoric, . . . . . 105
Kandura, *Gyrardinia heterophylla*, . . . 121
rubefacient, . . . . . . 105
Kanjika, gruel, . 49, 138, 140, 141, 142, 144, 145, 148
Kankayana, a sage, . . . . . . 30
Kansya, brass, . . . . . . 135
Kanthya, expectorant, . . . . . 105
Kapikachhu, *Mucuna pruriens*, . . . 105
Kapila, one of the five "Happy Sleepers," . . 79
Kapillaka, *Mallotus Phillippiensis*, . . . 129
Kapinjala, a sage, . . . . . . 30

PAGES
Karaba-dine-kaderi, a medical work, . . . 199
Karanja, *Pongamia glabra*, . . . . 58, 129
Karapatra, a surgical instrument. *See* Plate IX., . 183
Kardama, hydrous silicate of aluminum, . . 136
Kareera, *Capparis aphylla*, . . . . 56, 184
Karkoti, *Dregia volubilis*, . . . 14
Karma, action, . . . . 152, 153
Karnavedha, a ceremony, . . . . 54
Karpoora, *Camphora officinarum*, . . . 107
Karsha, a weight, . . . . 149, 150
Karshyakara, antifat, . . . . 106
Kartarika, a surgical instrument. *See* Plate IX., . 183
Kasa-ghna, *Terminalia bellerica*, . . 105, 112
Kasha, *Poa cynosuroides*, . . . . 106
Kashaya. *See* Astringent, . . . 103, 114
Kashiraj. *See* Devodas, . . . . 33, 34
Kashyapa, a sage, . . . . . 30
Kasisa, sulphate of iron, . . . . 135
Kasturi, musk, . . . . 64, 78, 81, 131
Kasumba, safflower, . . . . . . 120
Katakafala, *Strychnos potatorum* and *Strychnos nux vomica*, 129
Katala, a tree, . . . . . 59
Katfala, *Myrica sapida*, . . . . 129
Katu, pungent, . . . . 102, 114
Katuki. *See* Katurohini, . . . . 50, 104
Katukoshtaki, *Laffa amara*, . . . . 128
Katurina. *See* Bhoostrina, . . . 104, 128
Katurohini, *Picorrhiza kurroo*, . . . 50, 104
Katyayana, a sage, . . . . . 30
Kaumara-bhritya, pœdiatrics, . . . 25, 155
Kaundinya, a sage, . . . . . 30
Kauravas, descendants of Kuru, . . . 188
Kavalika, suppository, . . . . 148
Kayachikitsa, practice of physic, . . . 25
Keetamari, *Aristolochia bracteata*, . . . 122

PAGES

Kelale-shawshoor-al-Hindi, Arabic translation of Sushruta, 196
Kepler, . . . . . . . 169
Kerba, *Panitis succinifer*, . . . . 127
Kesara, *Crocus sativus*, . . . . 108
Kesha, hair, . . . . . . . 131
Keshava, father of Bopedeva, . . . 39
Ketaki, *Pandanus odoratissimus*, . . . 59
Ketu, an imaginary planet, . . . . . 169
Khakhas, poppy seeds, . . . . . 120
Khalif, a ruler, . . . . . . 196
Khari, a measure, . . . . . . 149
Kharjura, *Phœnix dactylifera*, . . . 127
Khatika, carbonate of calcium, . . . 136
Khorasani vacha, *Acorus Calamus*, . . . 38
Kidney bean, . . . . . . . 82
Kirata, *Ophelia Chiretta* and *Ophelia angustifolia*, 112, 129
Kishkindha, a town, . . . . . . 35
Kledana, a kind of Kafa, . . . . . 89
Knife, . . . . . . 177, 184
Koka, a writer on postural treatment, . . . 74
Kola, a weight, . . . . . . 149
Koorchas, glands, . . . . . . 97
Koorma, a vital air, . . . . . . 88
Krikala, a vital air, . . . . . . 88
Krimighna, anthelmintic, . . . . 106, 112
Krimikrit, vermiparous, . . . . . 106
Kripa, a sage, . . . . . . 58
Krishnayajur-veda, one of the Vedas, . . . 21
Kritayuga, the first age of the world, . . . 19, 35
Kritharika, a surgical instrument. *See* Plate X., . 183
Ksharagad, a pharmaceutical preparation, . . 156
Kshatriya, the fighting class, . . . . 16, 54
a kind of diamond, . . . . . 142
Kshaudra, a kind of honey, . . . . . 131
Ksheerakakoli, *Heydysarum gangeticum*, . . . 108

PAGES

Ksheerasagara, sea of milk, . . . . . 51
Kshirapani, a writer on medicine, . . . 29
Kubera, a malignant spirit, . . . . 159
Kudava, a measure, . . . . . 150
Kulatha, a kind of vetch, . . . 138, 141
Kumbhika, kind of impotent person, . . 47
Kunduru, *Boswellia thurifera*, . . . 123
Kurvinda, a stone used in surgery, . . . 184
Kushapatra, a surgical instrument. *See* Plate X., . 183
Kushika, a sage, . . . . . . 30
Kushtaghna, antiscorbutic, . . . . 106
Kushtha, *Saussurea lappa*, . . . . 108
Kvatha, decoction, . . . . . 148

L

Labour, . . . . . . 48, 151
Lac, sealing wax, . . . . . 131
Lachrymation, . . . . . 60
Lactiferous ducts, . . . . . 98
Lajjalu, *Mimosa sensitiva*, . . . . 111
Laksha, lac, . . . . . . 131
Lalaghna, antisialagogue, . . . . 106
Lalotpadaka, sialagogue, . . . . 106
Langakshi, a sage, . . . . . 30
Langali, *Gloriosa superba*, . . . . 49
*Langenaria vulgaris*, . . . . 185
Lanka, Ceylon, . . . . . 187
Lapis lazuli, . . . . . . 65, 135
Largus, Scribonius, . . . . . 127
Lataka Mishra, father of Bhava Mishra, . . 119
Lavana, salt, . . . . . . 102
Laxative, . . . . 102, 105, 116, 130, 131
Layers of the skin, . . . . . 97, 98
Lead, . . 15, 78, 134, 135, 139, 141, 142, 147
Leaves, betel, . . . . . . 69

PAGES
Leaves, plates of, . . . . . . 65
Leeches, . . . . . 131, 184
water, . . . . . . 154
Lekhana, liquefacient, . . . . 106
Lekhya, scarifying, . . . . . 181
Lemon, . . . . . 139, 145
Lepa, plaster, . . . . . . 148
Leprosy, . . 33, 117, 127, 134, 147, 152, 153, 168
Leucorrhœa, . . . . . . . 130
Life, . . . . . . 72, 190
duration of, . . . . . . 172
prolongation of, . . . 61, 63, 70, 146, 147
Ligaments, . . . . . . . 97
Ligaturing, . . . . . . . 185
Light, . . . . 69, 81, 83, 88, 190
Lime, . . . . . . 69, 136
Liniment, . . . . . . . 148
*Linum usitatissimum*, . . . . 129
Liquefacient, . . . . . . . 106
Liquids, . . . . . . . 66, 82
Liquorice, . . . . . . . 182
Litharge, . . . . . . . 135
Lithonlytic, . . . . . . . 104
Liver, . . . . 46, 89, 92, 140, 153
Lizard, bite of, . . . . . . 155
Load-stone, . . . . . . . 135
Lochana-hita, *Cassia Absus*, . . . 113, 136
Loha, iron, . . . 134, 135, 136, 141, 142, 147
Lohini and Lohita, a layer of the skin, . . 97, 98
Lolimbraja, a writer on medicine, . . . 39
Longevity, . . . . 57, 58, 65, 117, 147
Looking-glass, . . . . . . 58, 60
Lotion, . . . . . . . 148
Lozenge, . . . . . . . 148
Lucretius, . . . . . . . 75

PAGES
*Luffa amara*, . . . . . . 128
*echinata*, . . . . . 107
Lunatic, . . . . . . 165, 168
Lying-in-room, . . . . . 47

M

Macrobus, a writer on dreams, . . . 166
Mada, secretion from an elephant's temples, . . 131
Madaka, inebrient, . . . . . 106
Madan-us-shifa-i-sikandar, a medical treatise, . 199
Madana, *Randia dumetorum*, . . . 109
Madana Vinoda, a medical work, . . . 120
Madanpala, Raja, . . . . . 120
Madhava or Madhavacharya, a medical writer, 35, 36, 79
Madhavanidana, a work on medicine, . . 33, 36
Madhavavedartha prakasha, a work by Madhava, . 35
Madhavavritti, a work on grammar, . . . 35
Madhu, honey, . . . 38, 52, 131, 132, 181
Madhujama, wax, . . . . 132, 185
Madhuka, *Glycyrrhiza glabra*, . . . . 109
Madhura, sweet, property of, . . . 102
Madhusrava, *Bassia latifolia*, . . . 116
Madya, *Vina medicata*, . . . . 107
Magnetism, . . . . . . 203
Mahabharata, . . . . 187, 188, 194
Mahanamya, a religious rite, . . . 54
Mahomed Akbar Arzani, a physician, . . 199
Mahomed Momin, a physician, . . . 199
Mahomedan drugs introduced into India, . 126, 127
rule, decline of Hindoo medicine, . 126, 127, 198
Maitreya, a sage, . . . . . . 30
Majja, marrow, . . . . 46, 92, 96
Makshika, a kind of honey, . . . . 131
*Mallotus Phillipiensis*, . . . . 129
Mamanjaka, *Hipian orientali*, . . . 122

PAGES

Mammary glands, . . . . . 51, 96
Manas, mind, . . . 25, 41, 67, 70, 163
Manasshila, bisulphide of arsenic, . . 135, 141
Mandalagra, a surgical instrument. *See* Plate X., . 183
Mandura, hydrated oxide of iron, . . . 136
Manek, Bhatta, writer on medicine, . . . 39
Manes, . . . . . . . 74
Mangala, the planet Mars, . . . 69, 169, 170
Mango, . . . . . . 59, 66
Mania, dancing, . . . . . . 202
Manika, a measure, . . . . . . 149
Manjana, dentifrice, . . . . . . 148
Manjishtha, *Rubia cordifolia*, . . . 109
Mansa, flesh, . . . 70, 81, 92, 95, 132, 165
Mansadhara, a layer of the skin, . . . 98
Manu, the Hindoo law-giver, 12, 16, 20, 62, 63, 65, 66, 204
Manushi, herb-doctor, . . . . 204
Manu-vina, name of a king, . . . . 194
Marichi, a weight, . . . . . . 149
*Piper nigrum*, . . . . 105, 123
Markandeya, a sage, . . . . . 30, 58
Marmasthanas (vitals), . . . . 98
Marrow, . . . . . 46, 92, 96
Maruta, a malignant spirit, . . . 159
Masha, a weight, . . . . . . 150
*Phaseolus radiatus*, . . . . 81, 82
Mashaka, a weight, . . . . . . 149
Massage, . . . . 70, 71, 81, 82, 203
Materia medica, 101, 104, 113, 117, 122, 128, 135, 196, 203
Matha, *Phaseolus aconitifolius*, . . . 106
Matkuna, bed-bug, . . . . . . 131
Matter, . . . . 40, 41, 42, 43, 186
Mauktika, pearl, . . . . . 133, 136
Maund, a weight, . . . . . . 150
Mayofala, *Quercus infectoria*, . . . 107

PAGES
Meatus urinarius, . . . . . . 98
Medas, fat, . . . 70, 92, 95, 102, 109, 132
Medica, henna, . . . . . . 123
Medical science, Indian, 11, 75, 85, 117, 124, 168, 189, 191, 199, 200, 204, 205
  Western, . . . . 200, 203
Medicine, . . . 22, 59, 146, 151, 176, 190, 193
  cold-powered, . . . . . 108
  European, . . . . 196, 203, 205
  forms of, . . . . . 148
  Hindoo system of, . . 187, 193, 199, 201
  hot-powered, . . . . . 130
  Indian, . 188, 189, 190, 197, 198, 200, 202, 204, 205
  opening, . . . . . 70
  restorative, . . . . . 197
  writers on, . . . . . 77, 151
Megasthenes, . . . . . 84
*Melia Azadirachta*, . . . . . 49
Memory, . . . . . . 89, 139
Menorrhagia, . . . . . . 131
Menses, . . . . . . 96
*Mercurius Vitæ*, . . . . . 147
Mercury, . 38, 135, 136, 139, 143, 144, 145, 146, 147
  Bhasma of, . . . . . 145
Mesmer, . . . . . . 204
Messenger, to call a doctor, . . . 165, 166
Metals, used as medicines, 134, 135, 136, 137, 138, 147, 148
Methica, fenugreek, . . . . . 120
Mica, . . . . . . 135
*Michelia Champaca*, . . . . . 59
Microcosm, . . . . . . 41, 85
Micturition, . . . . . . 141
Midwife, . . . . . . . 47
Milk, 49, 50, 51, 63, 67, 78, 79, 82, 96, 102, 130, 146
Mimansa, ritual philosophy, . . . . 17

PAGES

*Mimosa Sirissa*, . . . . . 109
*sensitiva*, . . . . . 111
Mind, . . . . 25, 41, 67, 70, 163
Mineral, . . . . . 134, 155, 203
drugs, . . . . . 163, 202
kingdom, . . . . . 101
products, . . . . . 148
Mint, . . . . . . 123
Mishra-sthan, . . . . . 194
Mishra, temperate country, . . . . 56, 57
Misra, Egypt, . . . 32, 193, 194, 195
Mlechha, barbarian, . . . . . 194
Modaka, bolus, . . . . . 148
Molasses, . . . . . . 91, 165
Monier-Williams, Sir Monier, . . . 201
Moon, . . . . . 14, 169, 170
Moon-light, advantages of, . . . 73, 82, 83
Moon-stone, . . . . . . 136
Mootra, urine, . . . . . 133
Mootrasangrahana, anuretic, . . . . 106
Mootravirechaneeya, diuretic, . . 106, 177
Moreshvar, Bhatta, a writer on medicine, . . 39, 122
*Moringa pterygosperma*, . . . . 106
Mother's milk, qualities of, . . . . 49
Mouth, . . . . . 87, 98, 173, 181
Mrita-vatsa, a kind of barren woman, . . 47
Muchakanda, a sage, . . . . . 79
*Mucuna pruriens*, . . . . 105
Mucus, hypersecretion of, . . . . 60
nasal, . . . . . . . 96
Mudrika, a surgical instrument. *See* Plate X., . 183
Mukha-karana, a preparation of mercury, . . 145
Mukta, missive weapons, . . . . . 16
pearl, . . . . . . 133, 136
Muktamukta, both missive and non-missive weapons, 16

PAGES
Müller, Max, . . . . . 17
Muni, sage, . . . . . . 33
Murvya, *Sanseviera zeylanica*, . . . 50
Muscles, . . . . . . 95, 98
Musk, . . . . . 64, 78, 81, 131
Mustard seed, . . . . . 58, 144
*Myrica sapida*, . . . . . 129
*Myristica moschata*, . . . . . 106
Myrobalans, the three, . . . 178, 195
    belleric, . . . . 105, 112
    chebulic, . . . . 52, 114
    emblic, . . . . . . 63

N

Nabhi-yantra, an apparatus. *See* Plate IV., . . 144
Nadis, catheters, . . . . . 182
Naga, a vital air, . . . . . 88
Nail, . . . . 45, 96, 133, 184
Nairuti, a spirit, . . . . . 159
Nakha. Nail, . . . 45, 96, 133, 184
Nakha-shastra, a surgical instrument. *See* Plate X., 183
Nakula, one of the Pandavas, . . . 188
Nala, king of Nishadh, . . . . 188
Nama-karana, a ceremony, . . . . 53
Nandana, a physician, . . . . 121
Narada, a sage, . . . . . . 30
Narahari, a physician, . . . 120, 121
*Nardostachys jatamansi*, . . . 108, 123
*Narthex Asafœtida*, . . . 107, 142, 145
Nasya, snuff, . . . . . 148
*Nauclea Cadamba*, . . . . . 59
Nausea, . . . . . . 132
Navanita, butter, . . . 52, 58, 82, 130, 170
Navasadara, chloride of ammonium, . . . 136
Navatanka, a measure, . . . . 150

PAGES

Navel, . . . . . . 85, 89, 111
Nearchus, . . . . . . 189
Neck, bones of, . . . . . 96
*Nelumbium speciosum*, . . . . 107
Nervine tonic, . . . . . 141
Nervous system, . . . . 71, 202, 232
Nictation, . . . . . . 88
Nidan, diagnosis, . . . . . 151
Nidananjana, a treatise on diagnosis, . . 29
Nidrahara, anti-hypnotic, . . . . 106
Nidrakara, hypnotic, . . . . 106
Nighantaratnakara, a treatise on materia medica, . 122
Nighantu, dictionary, . . . . 15, 196
Nila, sapphire, . . . . . 135
Nila-shikhandi, a Tantric deity, . . . 193
Nila-tantra, a religious doctrine, . . . 194
Nimba, *Melia Azadirachta*, . . . . 50
Nirgundi, *Vitex Negundo*, . . . 110, 184
Niromakara, depilatory, . . . . 106
Nishkramana, a religious ceremony, . . . 54
Nitrates, . . . . . . 15, 136
Nitric Acid, . . . . . . 14
Noorudeen Mahomed Abdulla Shirazee, an author, . 199
Nose, . . . . . . 87, 102
  cartilages of, . . . . . 95
  drinking water through, . . . . 79
  formation of new, . . . . 178
Nostril, . . . . . 78, 98, 173
Nostrum, . . . . . . 125
Nubility, . . . . . . 45
Nut-meg, . . . . . . 69, 78
Nutritive, . . . . . . 92, 127
Nuts, Indian, . . . . . 124
Nyaya, logic, . . . . . 17

## O

PAGES
Obesity, . . . . . . 67, 139, 141
Oil, . . . 60, 81, 138, 140, 141, 148, 165, 181
rubbing of, . . . . . 61, 81.
Ointment, . . . . . 129, 132, 148
Oleaginous preparations, . . . . 137
Olfaction, . . . . . . . 156
Omens, . . . . . . . 165
*Onosma bracteatum*, . . . . . 127
Onyx, medicinal use of, . . . 135, 136
Operation, 48, 176, 178, 180, 181, 184, 186, 193, 197, 203
abdominal, . . . . . 181
about the mouth and anus, . . . 181
anæsthetics in, . . . . . 184
cataract, . . . . . 178
dexterity of the Indians in, . . . 180
obstetric, . . . . . 48, 178
ophthalmic, . . . . . 178
*Ophelia angustifolia*, . . . . . 129
*Chiretta*, . . . . • 112, 129
Ophthalmia, . . . . . . 33
Opium, . . . . . 120, 126, 127
Ordinances of Manu, . . . . 204
Organs of excretion, . . . . . 42
of generation, . . . . . 42, 74
of locomotion, . . . . . 42
pelvic, . . . . . . 75
of prehension, . . . . . 42
of sense, . . . . 41, 42, 73
Oviparous, . . . . . . 43
Oxidation of metals, . . . . . 137
Oxides, . . . . . 14, 15, 136
of iron, . . . . . . 15
of lead, . . . . . 15, 78, 135

## P

PAGES

Pachaka, a kind of Pitta, . . . . . 88
Pad, foot, . . . . . . 42
Padati, infantry, . . . . . 16
Padmaraga, ruby, . . . . . 135
Pain, . . . . 68, 79, 103, 126
Paka, confection, . . . . 148
Pakvashaya, small intestines, . . 88
Pala, a weight, . . . . 149, 150
Palandu, *Alium Cepa*, . . . . 128
Palasha, *Butea frondosa*, . . . . 128
Palpation, . . . . . . 156
Pamshujakshara. *See* Potassium, . . 136, 145
Pan, betel-leaf, . . . . 69, 78, 145
Panaka, syrup, . . . . . 148
Panasa, *Artocarpus integrifolia*, . . 109, 185
Pancha-dashi, a work on philosophy, . . 35
*Pandanus odoratissimus*, . . . . 59
Pandavas, princes of Hastinapura, . . . 188
Pandit, scholar, . . . 73, 122, 197
Pandu-roga, anæmia, . 62, 94, 133, 142, 153, 154, 168
Pani, hand, . . . . . . 42
*Panicum miliaceum*, . . . . . 107
Panini, the first grammarian, . . . 15, 20
*Panites succinifer*, . . . . . 127
Paracelsus, . . . . . 146, 147
Parada. Mercury, 38, 135, 136, 139, 143, 144, 145, 146, 147
Paralysis, . . . . . 132, 154, 203
Parashara, a law-giver, . . . . 16
writer of medicine, . . . . 29, 30
Parasika vacha, *Acorus Calamus* of Persia, . . 38
Parikshaka and Parshurama, a sage, . . . 30, 58
Parturifacient, . . . . . 107
Parturition, . . . . . . 47, 76
Paseer, a weight, . . . . . 150

PAGES
Pashanabheda, carbonate of iron and lime, . . 136
Pashanabhedi, *Coleus aromaticus*, . . . 112
Paste, . . . . . 48, 58, 134, 148
Patala-yantra, an apparatus. *See* Plate IV., . 144
Patanjala, a sage, . . . . . . 76
Patha, *Cissampelos glabra*, . . . . 50
Pathology, . . . . . . . 99, 202
Patol, *Trichosanthes dioica*, . . . . 50
Paushkalavata, a writer on surgery, . . . 176
Pautika, a kind of honey, . . . . 132
Payu, organ of excretion, . . . . 42
Pearl, medicinal use of, . . . 133, 136
Peeloo, *Salvadora indica*, . . . . 56, 105
Pellitory, . . . . . . . 120
Pepper, . . . . . . 52, 58, 145
Perception, . . . . . . . 94
Percussion, . . . . . . . 156
Persian drugs, . . . . 11, 122, 126
Perspiration, . . . . . . . 102
Peruka, guava, . . . . . . 123
Peshi, muscle, . . . . . . 95
Peshwas the, Maratha rulers of Poona, . 199, 200
Peya, emulsion, . . . . . . 148
Pharmacodynamics, native, . . . . 115
Pharmacy, Indian, . . . . . 128, 198
*Phaseolus radiatus*, . . . . . 81, 82
*aconitifolius*, . . . . . . 106
Phlegm, . 70, 81, 85, 86, 96, 97, 113, 115, 157, 191
removal of, . . . . . 59, 69, 81
Phlegmatic, . . . . . . 56, 57, 91
Phlegmagogue, . . . . . . 69, 105
*Phœnix dactylifera*, . . . . . 127
Phthisis, . . . . . . . 130
*Physalis somnifera*, . . . . . . 109
Physician. *See* Vaidya, 103, 122, 134, 161, 170, 179, 188, 204

PAGES

Physician—*continued*.
court, . . . . . 187, 199
duties of, . . . 57, 161, 162, 164, 187
fees of, . . . . . . 162
qualities of, . . . . . 161
Physiology, . . . . . 93, 179, 203
Pichha, feather, medicinal use of, . . . 133
*Picorrhiza kurroo*, . . . . . 50, 104
Piles, . . . . 79, 133, 154, 178, 203
amulet for curing, . . . . 160
Pill, . . . . . . 137, 148
*Pimpinella Anisum*, . . . . 109, 129
Pimples, . . . . . . 59, 97
Pincers, . . . . . . 182
Pindi, poultice, . . . . 148, 177, 181
Pine-apple, . . . . . . 123
Pinjrapole, animal hospital, . . . . 189
*Pinus deodara*, . . . . . 50
*Piper longum*, . . . . . 105
*nigrum*, . . . . . 105, 123
Pippala, *Ficus religiosa*, . . . . 59, 106
Pippalachhala, *Cortex Ficus religiosa*, . . 106
Pippalimoola, *Piper longum*, . . . . 105
Pishacha, a demon, . . . . . 160
Pistachio nut, . . . . . . 120
Pistam, Pistachio, . . . . . 120
Pitta. *See* Bile, . . . . 85, 86, 88, 133
Pittahara, anticholeric, . . . . 107
Pittakara, cholagogue, . . . . 107
*Plantago Ispaghula*, . . . . . 122
Plants, . . . . . . . 80
nomenclature of Indian, . . . . 110
Plaster, . . . . . . 148, 177
Plato, . . . . . . 115, 191
Playfair, . . . . . . . 14

PAGE
Pleeha-ghnee, *Thevetia neriifolia*, . . . 112
Plexuses, . . . . . . 97
Plota, lotion, . . . . . 148
*Plumbago zeylanica.* *See* Chitraka, . . 116, 144, 177
*Poa cynosuroides*, . . . . . 106
Pocock, Mr, . . . . . . 191
Poisons, . . . . . 65, 155, 156
Pomegranate, . . . . 59, 66, 104
*Pongamia glabra*, . . . . . 58, 129
Poosha, Indra, . . . 26, 27, 28, 29, 31
Poppy-seeds, . . . . . 120
Portuguese disease, . . . . . 37
Posture, treatment by, . . 75, 76, 147, 204
Potassium, carbonate of, . . . 136, 145
nitrate of, . . . . . 136
Poultice, . . . . . 148, 177, 181
Powder, . . . . . 58, 140, 148
Prabhava, nature of a drug, . . . 101, 116, 117
Prajasthapana, anecbolic, . . . . 107
Prakriti, matter, . . . . 40, 41, 42, 43
Pramathi, antiphysical, . . . . 107
Prana, one of the vital airs, . . . . 66, 87
Prasavaka, parturifacient, . . . . 107
Prasruti and Prastha, a measure, . . . 149
Prativasa, antibromic, . . . . 107
Pravala, coral, . . . . 133, 136, 170
Pre-budhistic, . . . . . 186
Precious stone, . . . 64, 134, 135, 142
Pregnancy, . . . . . 46, 47, 96, 152
*Premna serratifolia*, . . . . . 108
Principles and practice of medicine, . . . 36
Prithvi, earth, . . . . . . 42
Priyangu, *Panicum miliaceum*, . . . 107
Probing, . . . . . 181, 184, 185
Proctor, Mr, . . . . . . 169

PAGES

Prognosis, . . . . . . 164, 203
Property, . . . . . . 80, 110, 142
Prophylactic, . . . . . 82, 117
*Prunus bocariensis*, . . . . . . 127
Ptolemy, . . . . . . . 14
*Ptychotis ajowan*, . . . . . 108, 129
Pudina, mint, . . . . . . 123
Puking, . . . . . . . 51
Pulastya, a writer on medicine, . . . 30
Pulmonary catarrh, . . . . . . 112
Pulse, . . . . . 156, 157, 158, 173
Punarnava, *Boerhaavia diffusa*, . . 109, 140
Puncturing, . . . . . . . 181
Pungent, . 66, 68, 81, 82, 83, 101, 102, 103, 114, 115
*Punica granatum*, . . . . . 59, 66, 104
Punsavana, a religious rite, . . . . . 53
Punya, merit, . . . . . . 162
Purans, sacred works of the Hindoos, . . . 194
Purgatives, . . . 83, 104, 107, 116, 155, 164
Purification of metals, . 137, 138, 139, 140, 141
of diamond and other precious stones, . . 142
of mercury, . . . . . . 143
Purisha, dung, medicinal use of, . . 134, 140, 142
Purishasangrahana, astringent, . . . 107
Purudhyaksha, magistrate, . . . . . 21
Purusha, a kind of diamond, . . . . 143
spirit, . . . . . . 40, 42, 43
Pus, . . . . . . . . 186
Pusha, a Vedic deity, . . . . . 29
Pushpa-phala, *Cucurbita maxima*, . . . 185
Pushpa-raga, topaz, . . . . . . 135
Pustules, . . . . . . 154, 179
Pyrites, . . . . . . . 135
Pythagoras, . . . . . . 14, 115, 191

## Q

PAGES
Quackery, . . . . . . 200
Qualities, of the five elements, . . . 115
Quartan fever, cause of, . . . . 131
Quenching of thirst, . . . . . 67
*Quercus infectoria*, . . . . . 107
Quicksilver, potion of, . . . 146, 147
Quieting a weeping child, an amulet for, . . 160

## R

Rahu, an obscure planet, . . . . 169
Rajanighanta, a work on Materia Medica, . . 120
Rajas, germ, . . . . . . 46
Rajataru, *Cassia fistula*, . . . 109, 111
Rajavarta, lapis lazuli, . . . . 65, 135
Rajika, a measure, . . . . . 149
Rajjus, chords, . . . . . 97
Rakshasi, demoniacal, . . . . 204
Rakta. *See* blood, . . . . . 92
Rala, *Shorea robusta*, . . . . 106, 129
Rama, King of Ayodhya, . . . . 33, 187
Ramayana, . . . . . . 187
*Randia dumetorum*, . . . . . 109
Ranjaka, a kind of Pitta, . . . . 89
Rapeseed, . . . . . . 60
Rasa, a mineral substance, . . . 134, 135, 146
  nourishing juice, . . 88, 92, 93, 94, 96
  taste, . . . . . 41, 88, 92, 101
Rasaka, carbonate of zinc, . . . . 135
Rasakarpura, corrosive sublimate, . . . 136
Rasana, a kind of Kafa, . . . . 90
Rasapradipa, a medical work, . . . 89
Rasasindura, sulphide of mercury, . . 135, 136
Rasayana, rejuvenescent, . . . . 107
  treatment of medicines restoring youth, etc., . 26

PAGES
Raseshism, a religious doctrine, . . . 146
Ratharudha, car-fighter, . . . . 16
Rati, a weight, . . . . . 150
Ratnas, gems, . . . 33, 135, 136, 196
Raupya, silver, used as medicine, . . 64, 134, 139
Ravana, king of Lanka, . . . . 187
Razes, . . . . . . 34
Rechana, hydragogue, . . . . 107
Reeti, calcined zinc, . . . . . 135
Refrigerant, . . . . . 108, 130
Regimen, . . . . . . 49, 182
Rejuvenescent, . . . . . 107
Respiratory system, . . . . . 87, 202
Restorative, . . . . . 101, 107, 108
Retas, fecundating fluid, . . . . 46
Rhazes, Al Rasi, . . . . . 195
Rheumatism, . . 127, 130, 132, 140, 155
Rhinoplasty, . . . . . . 178
Rhubarb, . . . . . . 124
Rice, . . . . . 81, 82, 114, 146
Rice-husk, . . . . . . 144
*Ricinus communis*, . . . . 111, 129
Rig Veda, . . 19, 21, 22, 23, 24, 26, 28, 35, 191
Ring-worm, . . . . . . 133
Rishabhaka, *Helekteres isora*, . . . 108
Rites, religious, . . . . . 50, 53
Rock salt, . . . . . . 65
Rohana, epulotic, . . . . . 107
Romans, . . . . . 70, 143, 171
Roopa, form, . . . . . . 41
Rubefacient, . . . . . . 105
*Rubia cordifolia*, . . . . . 109
Ruby, . . . . . . 135
Ruchi, a king, . . . . . 28
Rudanti, *Cressa cretica*, . . . . 121

PAGES

Rudra, a divinity, . . . . . 28, 193
Rudraksha, *Elæocarpus ganitrus*, . . . 121

S

*Saccharum officinarum*, . . . 105, 108
Sadhaka, a kind of Pitta, . . . . 89
Sadhya, curable, . . . . . 164
Safflower, . . . . . . 120
Saffron, . . . . 63, 64, 78, 81, 108
Sahadevi, *Vernonia cinerea*, . . . . 116
Sal ammoniac, . . . . . 145
Saline substances, . . . . . 82, 114
Salt, 58, 65, 66, 68, 73, 81, 82, 89, 101, 102, 134, 136, 145, 181
*Salvadora indica*, . . . . . 56, 105
*Salvinia cucullata*, . . . . . 129
Samana, one of the vital airs, . . . 87
Samavartana, a ceremony, . . . . 54
Sama Veda, . . . . . . 19
Sammohana, anæsthetic, . . . 107, 197, 198
Sammohini, an anæsthetic agent, . . . 197, 198
Samshodhana, emetic and purgative, . . . 107
Sand, . . . . . . 136
Sandal, . . . . . 63, 105, 129
Sandal-wood, . . . . . 83
Sandasha, tongs, . . . . . 182
Sandhi, joint, . . . . . 96
Sanjeevasthapana, restorative, . . . 107
Sanjivini, a restorative medicine, . . 197, 198
Sankhya, discriminative philosophy, . . . 17
a sage, . . . . . . 30
Sankochana, constringent, . . . . 107
Sankranti, passage of a planet, . . . 76
Sankritya, a sage, . . . . . 30
*Sanseviera zeylanica*, . . . . . 50
Sanskaras, rites, . . . . . 53

PAGES

*Saussurea lappa*, . . . . . . 108
*Santalum album* and *flavum*, . . . 63, 105, 129
Santanika, cream, . . . . . . 130
Sapphire, medicinal use of, . . . . . 135
*Sapindus emarginatus*, . . . . . 111
Saptaparna, *Alstonia scholaris*, . . . 128
Sarayati, a king, . . . . . . 27, 28
Sarsapa, mustard, . . . . . 58, 144
a small weight, . . . . . 149, 150
Sarvadarshanasangraha, a work on Indian philosophy, 35
Satva, extract, . . . . . . . 148
Satvikas, possessing quality of goodness, . . 68
Sayana, the commentator, . . . . . 35
Scarification, . . . . . . 181, 188
Science of Aryan medicine, . 23, 24, 99, 192, 193, 202
decline of, . . . . . . 198
rise of, . . . . . . 123, 204
veterinary, . . . . . . 188
western, . . . 100, 201, 202, 204
Scorpion, . . . . . . 47, 155
Scribonius Largus, . . . . . . 127
Scurvy, . . . . . . . 131
Sechana, spray, . . . . . . 148
Sedative, . . . . . . . 110
Seer, a weight, . . . . . . 150
Seevanis, sutures, . . . . . 97, 181, 185
Semen, 44, 45, 46, 69, 87, 92, 96, 102, 108, 130, 152, 174
*Semicarpus Anacardium*, . . . . 105, 112
Sensibility, . . . . . . . 197
Serapion, . . . . . . 34, 195
Serpent, . . . . . 32, 33, 47, 131, 155
*Sesamum indicum*, . . . . . 81, 107, 123
Sfatika, alum, . . . . . . 135
Shabda, sound, . . . . . . 41
Shah Jehan, the Emperor, . . . . . 199

Shakuneya, a sage, . . . . . . 30
Shalakas, bougies, . . . . . . 182
Shalakya, treatment of diseases of parts above the clavicles, 25
Shalmali, *Bombax malabaricum*, . . . 185
Shalya, surgery, . . . . . . 25, 176
Shamaneeya, calmative, . . . . . 180
Shami, *Acacia farnesiana*, . . . . . 56
Shampooing, . . 70, 71, 81, 82, 89, 186, 204
Shana, a weight, . . . . . 149, 150
Shandha, a kind of impotent person, . . 47
Shandilya, a sage, . . . . . . 30
Shankara-digvijaya, life of Shankara, . . 36
Shankavisha, arsenious acid, . . . . . 137
Shankha, conch, . . . . . . 134
Shankhapushpi, *Clitoria Ternatea*, . . . 53, 110
Sharad, the sultry season, . . . 80, 82, 83
*Sharaka indianus*. *See* Charaka, . . . 195
Sharaloma, a sage, . . . . . . 30
Shararimukha, a surgical instrument. *See* Plate X., 183
Sharira, physical disease, . . . . . 55
Sharngdhara, a writer on medicine, . 39, 126, 144, 152
Shastra, book of religion, . . . . 37, 40
Shastras, surgical instruments, Plates IX., X., . 182
Shatapushpa, *Pimpinella Anisum*, . . 109, 129
Shatavari, *Asparagus tomentosus*, . . . 53
Shathi, *Hedychium spicatum*, . . . 105
Shaunaka, a sage, . . . . . . 30
Sheetaprashamana, antalgide, . . . 108
Sheeta-veerya, cooling, . . . 113, 114, 130
Shefali, *Vitex Negundo*, . . . 110, 184
Shesha, serpent-god, . . . . . . 32
Shigru, horse-radish tree, . . . . . 145
Shigrubeeja, *Moringa pterygosperma*, . . 106
Shilajita, bitumen, . . . . . . 135
Shireesha, *Acacia Serissa*, . . . . 59, 109

PAGES

Shirovirechana, sternutatory, . . . 108
Shishira, dry season, . . . . 80, 81, 83
Shleshana, a kind of Kafa, . . . . 90
Shodhala, a writer on materia medica, . . 121
Shofakara, epispastic, . . . . 108
Shoolprashamana, anticolic, . . . . 108
Shooting-pain, . . . . . 79, 103
*Shorea robusta*, . . . . 106, 129
Shotha-ghnee, *Trianthema obcordata*, . . 112
Shothahara, discutient, . . . . 108
Shoulder-joint, . . . . . 90
Shramahara, refrigerant, . . . 108, 130
Shravana, a Hindoo month, . . . . 32
Shringa, horn, . . . . . 134
Shringavera, *Zingiber officinale*, . . . 123
Shronitasthapana, styptic, . . . . 108
Shrotra, organ of hearing, . . . . 41
Shrutambu, water in which hot iron is quenched, . 148
Shudra, a kind of diamond, . . . . 142
Shuka, water-leeches, . . . . 154
Shukra, semen, . . . . . 92, 96
Shukrajanana, spermatopoietic, . . . 108
Shukrala, tonic, . . . . . 108
Shukrashodhana, semen-improver, . . . 108
Shukta, ferment, . . . . . 148
Shurana, *Arum Colocasia*, . . . . 140
Shvasahara, antasthmatic, . . . . 108
Shveta, a layer of the skin, . . . . 97
Sialagogue, . . . . . . 106
Siddha, hypnotist, . . . . . 204
Sight. *See* Vision, . . . 41, 60, 62, 68, 89
Sikander, Alexander the Great, . . 15, 189, 190
Sikata, silica, . . . . . 136
Silver, Bhasma, . . . . . 139
medicinal use of, . . . . 64, 134

PAGES
Simantas, supporters, . . . . . 97
Simantonnayana, a ceremony, . . . 53
Sindhava, chloride of sodium, . . . 136
Sindhi Chara, a term applied to Vagbhata, . . 196
Sindura, red oxide of lead, . . . . 78, 135
Sisaka. *See* Lead, . . . . . . 134
Bhasma, . . . . . . 141
Sitaphala, custard-apple, . . . . . 123
Sivya, suturing, . . . . . . 181
Skin, . 62, 70, 72, 89, 103, 130, 131, 153, 156, 185
layers of, . . . . . . 97
to improve colour of, . . . . . 61
Skin-flaps, transplanting of, . . . . 178
Sleep, . 55, 61, 70, 71, 72, 73, 78, 79, 81, 82, 127, 130
Sleepers, "The Five Happy," . . . 79
Small-pox, . . . . . . . 179
Smell, organ of, . . . . . . 42
Smith, Dr, . . . . . . . 32
Snake-bite, . . . 133, 152, 155, 184, 189
Snayu, a ligament, . . . . . . 97
Sneezing, . . . . . . . 88, 174
Snehana, a kind of Kafa, . . . . . 90
Snehopaga, demulcent, . . . 109, 127, 131
Snoohee, *Euphorbia Tirucalli*, . . . 108
Snuff, . . . . . . . 148
Soda, carbonate of, . . . . . . 136
Sodium, chloride of, . . . . . . 136
Sohamji, Vaidya, . . . . . . 122
*Solanum indicum*, . . . . . 105, 142
Solvents, . . . . . . . 177
Soma, a beverage, . . . . . . 28
Sonamukhi, *Cassia lanceolata*, . . . 128
Sores, . . . . . . . 154
Soul, . . . . . . . 42, 46
Sound, . . . . . . . 41

PAGES

Sour, . . 68, 82, 83, 101, 102, 114, 138, 140
Sparsha, touch, . . . . . . 41
Specific gravity, . . . . . . 93
Sperm, . . . . . . . 44
Spermatopoietic, . . . . . . 108
Sphatika, crystal, . . . . . . 136
Spider, bite of venomous, . . . . 155
*Spilanthes oleracea*, . . . . . 38, 106
Spirit, . . . . 17, 18, 40, 41, 66
  evil, . . . . . . 64, 159
Spleen, . . . . 46, 89, 92, 134, 140
Sprains, . . . . . . 134
Spray, . . . . . . 148
Sransana, drastic, . . . . . 109
Srotanjana, sulphide of lead, . . . 135
Stanyajanana, galactagogue, . . . . 109
Sterno-clavicular joints, . . . . 90
Sternum, . . . . . . 87
Sternutatory, . . . . . . 108
Sthaulyakara, promoting fat, . . . 109
Sthavara, a kind of poison, . . . 155, 156
Stimulant, . . . . . 69, 127, 139
Stomach, . . . 67, 72, 86, 87, 88, 89, 114
Stomachic, . . . . . . 105
Stone, . . . . . 65, 117, 155
  dissolving of, . . . . 112, 177
  precious, . . . 64, 134, 135, 142
Strength, promotion of, . . . 61, 102, 139
Stricture, . . . . . . 132
*Strychnos potatorum*, *nux vomica*, etc., . . 129
Styptic, . . . . . . 108
Sublimate, corrosive, . . . . 137
Suchi, a surgical instrument. *See* Plate X., . 183
Sugandhi, a kind of impotent person, . . 47
Sugar-cane, . . . . . . 66, 82

PAGES

Sukanya, a damsel, . . . . . . 27
Sulemani Kharjura, date fruit of Suleman, . . 38
Sulphates, . . . . . . 15, 135
Sulphur, . . . 135, 139, 140, 146, 147
Sulphuric acid, . . . . . . 14
Sun, the, . . 26, 54, 61, 62, 76, 81, 144, 169
its influence on human body, . . . 82, 168
sons of, . . . . . . . 190
Suppository, . . . . . . 148
Surgeon, . . 177, 178, 180, 181, 183, 187, 203, 204
qualifications of a, . . . . . 181
Surgery, 23, 25, 27, 33, 36, 176, 177, 182, 185, 186, 203
abdominal, . . . . . . 198
cranial, . . . . . . . 198
decline of Indian, . . . 185, 188, 204
modern, . . . . 177, 181, 198
plastic, . . . . . . . 178
Surgical instruments, . . . . . 25
Surpa, a weight, . . . . . . 149
Surya, the sun, . . . . . . 69
Suryakanta, sun-stone, . . . . . 136
Suryakshara, nitrate of potash, . . . . 136
Suryanilokana, a ceremony, . . . . . 54
Sushena, army surgeon of Rama, . . . . 187
Sushruta, a writer on medicine and his work, 32, 36, 51, 67
77, 94, 103, 118, 151, 163, 172, 176, 184, 193
a guide in surgery, . . . . . 176
Arabic translation of, . . . . . 196
classification of surgical operations by, . 180, 185
directions in operations by, . . 181, 182, 183
Sutrasthana, institutes of medicine, . . . 151
Sutures, . . . . . . 97, 181, 185
Suvarna, gold, medicinal use of, . . 134, 138
Suvarnamakshika, yellow pyrites, . . . 135
Svabhavika, natural, . . . . . . 55

PAGES
Svapneshvari, goddess of dreams, . . . 168
Svarasa, expressed juice, . . . . 148
Svarya, good for voice, . . . . 109
Svastika, forceps, . . . . . 182
Sveda, vapour, . . . . . 148
Svedaja, a subdivision of animal kingdom, . . 43
Svedopaga, diaphoretic, . . . . 109
Sweat, . . . . . . 88, 96, 113
Sweet, . . 66, 81, 82, 89, 91, 101, 102, 114, 115
Swelling, . . . . . 112, 155, 177
Swimming, . . . . . . 95
Symptomatology, Indian, . . . . 203
Syphilis, . . . . . . 37, 127
Syrup, . . . . . . 148

T

Tagara, *Valeriana Hardwickii*, . . . 129
Taila. *See* Oil, . . . . . 148
Takra, whey, . . . 67, 130, 138, 140
Talaka, tersulphide of arsenic, . . 135, 140
Talas, tongs, . . . . . . 182
Tamaku, tobacco, . . . . . 123
Tamarind, . . . . . . 140
Tamra, copper, medicinal use of, . . 134, 139
a layer of the skin, . . . . 98
Tankana, borax, . . . . . 135
Tantras, parts, . . . . . 25
Tantric, relating to the mystical science of the Tantras, 193
Tapta-khalva-yantra, Nos., I., II., III., Plate V., . 180
Taramakshika, white pyrites, . . . 135
Taste, . . . . 41, 90, 101, 102, 114
Tattvas, principles, . . . . . 41
Tea, . . . . . . . 122
Tears, . . . . . . . 96
Tejas, fire, . . . . . . 42

PAGES
Tendons, . . . . . . 46
*Terminalia bellerica*, . . . 105, 112, 113
  *Chebula*, . . . . 52, 104, 114
  *emblica*, . . . . . 63
Therapeutics, . . . . 36, 153, 202
*Thevetia neriifolia*, . . . . . 112
Thirst, . . . . . 55, 67, 102, 113
  diseases from excessive, . . . . 154
  means of allaying, . . . . 147
Throat, . . . . . . 65, 90
Tikta. Bitter, . . . . . 102
Tila, *Sesamum indicum*, . . . 81, 107, 123
Tin, medicinal use of, . . 15, 134, 140, 147
Tincture, . . . . . 137, 148
*Tinospora cordifolia*, . . . . 50, 52, 129
Tobacco, . . . . . . 58, 123
Tola, a weight, . . . . 149, 150
Tongs, . . . . . . 182
Tongue, . . . . . . 90, 156
Tonic, . . . 102, 108, 117, 130, 139, 141
Topaz, . . . . . . 135
Touch, . . . . . . 41
Toxic, . . . . . . 109
Toxicologist, . . . . . 189, 203
Trapu, *Cucumis pubescens*, . . . . 185
Trasrenu, a weight, . . . . . 149
Treacle, . . . . . . 52
Treatment, antiphlogistic, . . . . 155
  postural, . . . . . 75, 204
Trephine, . . . . . . 197
Treta-yuga, the second age of the world, . . 19
Triangles, double, to hasten delivery, . . . 48
*Trianthema obcordata*, . . . . 112
*Tribulus terrestris*, . . . . 104, 111
*Tricholapis glaberrima*, . . . . 121

PAGES

*Trichosanthes dioica*, . . . . . 50
Triconamiti, trigonometry, . . . . 14
Trifala, the three myrobalans, . . 178, 195
Trik, sterno-clavicular joints, . . . 90
Trikantaka, *Tribulus terrestris*, . . 104, 111
Trikurchaka, a surgical instrument. *See* Plate X., . 183
Trishnanigrahana, frigorific, . . . . 109
Trivrita, *Ipomœa Terpethum*, . . . 107
*Triumfetta angulata*, . . . . 121
Trunk, . . . . . . . 96, 97
Tuhfat-ul-Muminim, a medical treatise, . . 199
Tula, a measure, . . . . . 149
Tumour, . . . . 67, 98, 151, 155
    abdominal, . . . . 151, 154, 168
    method of treating, . . . . 180
    scrotal, . . . . . 151, 155
Turquoise, medicinal use of, . . . . 136
Tuttha, sulphate of copper, . . . . 135
Tvak. *See* Skin, . . . . . 130
    *Cinamomum Cassia*, . . . . 107

U

Udana, a vital air, . . . . . 87
Udbhija, sprouting, . . . . . 43
Udvartana, liniment, . . . . 148
Ugra-gandha, *Acorus Calamus*, . . . 110
Ulcer, varieties and treatment, . . 155, 180
Umbilicus, . . . . . 85, 89, 111
Undurkarnika, *Salvinia cucullata*, . . . 129
Universe, evolution of, . . . 18, 23, 99
Unnatural birth and labour, . . . 47, 151, 155
Upanaha, fomentation, . . . . 148
Upanayana, thread ceremony, . . . 54
Upa-rasa, secondary Rasa, . . . . 135
Upastha, organ of generation, . . . 42, 74

PAGE

Upa-veda, supplementary Veda, . . . . 23
Upayantras, dressings, . . . . . 182
Urdhvanalika, an apparatus, . . . . 144
Urethral discharges and sores, . . 151, 154
Urinary calculi, . . . . 151, 155, 177
 disorders, . . . . 141, 155
Urine, . . 87, 96, 153, 156, 167, 174
 as a remedy, . . . . . 133
 cow's, . . . . 133, 138, 140, 141
 dreams caused by abnormal secretion of, . 167
 examination of, . . . . . . 156
 retention of, . . . . . . 153
Ushira, *Andropogon nardus*, . . . 105, 128
Ushna-veerya. *See* Heating, . . . 113
Utkantaka, *Echinops echinatus*, . . . 122
Utpalapatraka, a surgical instrument. *See* Plate X., 183

Vaccination, . . . . . . . 179
Vacha, *Acorus Calamus*, . . 38, 53, 106, 110
Vadavanala, submarine fire, . . . . 68
Vagbhata, a writer on medicine, 34–36, 67, 90, 126, 188, 196
 classification of herbs by, . . . 119
Vagina, . . . . . . . 98
Vaidhrata, 27th astronomical yoga, . . . 76
Vaidurya, cat's-eye, . . . . . 136
Vaidya. Physician, 33, 83, 122, 125, 135, 143, 157, 163, 170, 187
Vaidya-jeevana, a medical work, . . . 39
Vaidyamrita, a work on medicine, . . . 39, 122
Vaidyavallabha, a medical treatise, . . . 39
Vaijepaya, a sage, . . . . . 30
Vaikhanasa, a sage, . . . . . 30
Vaisheshika, atomic philosophy, . . . 17
Vaishya, a kind of diamond, . . . . 142
 trading class, . . . . . . 54

PAGES
Vajeekara, aphrodisiac, . . .. 69, 109, 139
Vajikarana, means of increasing the virile power, . 26
Vak, organ of speech, . . . . 42
Vala, a weight, . . . .. . 150
*Valeriana Hardwickii*, . . . . 129
Vallaka, *Dolichos sinensis*, . . . . 109
Valmiki, author of Ramayana, . . . 187
Valukayantra, an apparatus. *See* Plate V., . . 144
Vamadeva, a sage, . . . . . 30
Vamana, emetic, . . . 81, 107, 109, 155
Vapour, . . . . . . 148
Varataka, cowry, . . . . . 134
Varnya, cosmetic, . . . . . 109
Varsha, the rainy season, . . . . 80, 82
Vartika, bougie, . . . . 148, 182
Varuna, a demon, . . . . . 159
Vasanta, spring, . . . . . 80, 81
Vasishtha, a sage, . . . . . 30
Vata, *Ficus indica*, . . . . . 58
*See* Wind, . . . . 85, 86, 87
Vatakara, promoting wind, . . . . 109
Vata-vairee, *Amygdalus communis*, . . . 111
Vatika, lozenge, . . . . . 148
Vatsanabha, *Aconitum ferox*, *A. Napellus*, &c., 109, 110, 128
Vatsayana, a sage, . . . . . 74
Vayu, air, . . . . . 41, 42, 94
Veda, . . . 17, 18, 19, 20, 22, 23, 35
antiquity of, . . . . . 19, 20
Vedanasthapana, anodyne, . . . . 109
Vedhya, puncturing, . . . . . 181
Vedic, civilization, . . . . 13, 21, 22
period, . . . . . 21, 33, 180
Vedini, a layer of the skin, . . . . 98
Veerya, power of a drug, . . . 101, 113
Vegetable, . . . . . . 80, 155

PAGES

Vegetable—*continued.*
kingdom, . . . . . . 101
products, . . . . 137, 147, 203
Veins, . . . . . . 46, 94, 183
Venery, . . . . . . 91
Venesection, . . . . . . 185
Venom, . . . . 124, 130, 133, 155
Vermifuge, . . . . . . 141
Vermiparous, . . . . . 106
*Vernonia cinerea*, . . . . . 116
Vessels, . . . . . 97, 150, 185
Vetas-patra, a surgical instrument. *See* Plate X., . 183
Vetch, . . . . . . 138
Vibhishana, a king, . . . . . 58
Vidanga, *Embelia ribes*, . . . 53, 106, 112
Vidari, *Batatas paniculata*, . . . . 109
Vidaringandha, *Costus speciosus*, . . . 104
Vidyapati, a writer on medicine, . . . 39
Vidyaranya, Madhava, . . . . 36
Vigour, restoration of, . . . . 139
Vikrama, king of Oojein, . . . . 196
*Vina medicata*, . . . . . 107
Vinegar, . . . . . . 148
*Viola odorata*, . . . . . 127
Vipaka, consequence of action of a drug, 101, 114, 115, 116
Vira Bukka, a king, . . . . . 35
Virile debility, . . . . . 155
power, . . . . . . 26, 102
Viscera, . . . . . . 96
Visha, toxic, . . . . . . 109
Vishaghna, antitoxic, . . . . 110
Vishnoo, a deity, . . . . . 57
Vishnu Vasudev Godbole, a Pandit, . . . 122
Vishnukranta, *Evolvulus hirsutus*, . . . 107
Vishudha, father of Charaka, . . . 33

PAGES

Vishvamitra, a sage and king, . . 30, 33, 194
Vision, . . . . . 41, 60, 62, 68, 89
Visravaniya, evacuating fluids, . . 181, 185
Vital airs, . . . . . 87, 88, 94
Vitals, . . . . . . 98, 183
*Vitex Negundo*, . . . . 110, 184
*Vitis vinifera*, . . . . . 116
Vivaha, marriage ceremony, . . . . 55
Viviparous, . . . . . . 43
Voice, . . . . 60, 69, 109, 132, 139
    to diagnose from, . . . . 156
Vomiting, . . . . . 62, 131, 155
Vriddhipatra, a surgical instrument. *See* Plate X., . 183
Vridha-trayi, the three old writers on Hindoo medicine, 35
Vrihimukha, a surgical instrument. *See* Plate X., . 183
Vullers, . . . . . . . 34
Vyaghri, *Solanum indicum*, . . . 105, 142
Vyana, a vital air, . . . . 87, 88, 94
Vyasa, a sage, . . . . . . 58
Vyatipata, an astrological yoga, . . . 76
Vyavahara Madhava, a work on Hindoo law, . 36
Vyayayi, sedative, . . . . . 110

W

Warts, . . . . . . . 155
Water, . . . 41, 42, 67, 82, 115, 168, 181
    an element, . . . . . 42, 115
    medicated, . . . . . 188
    medicinal properties of, . . . . 204
    one of the humours, . . . . 191
Water-leeches, diseases caused by, . . . 154
Watery extract, . . . . . . 148
Wax, . . . . . . 132, 185
Weakness, . . . . . . . 139
Weber, Dr, . . . . . . 178, 192

PAGES
Weights and measures, . . . 148, 149, 150
Wet-nurse, . . . . . . 50, 51
Wheat, . . . . . . 81, 82
Whey, . . . . 67, 130, 138, 140
White pyrites, . . . . . 135
Wilford, Major, . . . . . 194
Wilson, . . . . . . . 20
Wind, . . 69, 70, 85, 88, 91, 95, 109, 112, 157
affections, . . 56, 59, 65, 82, 90, 101, 103
dreams caused by, . . . . 167
an element, . . . . . 115
humour, . . . . 85, 86, 101
vessels for, . . . . . 96, 97
Wise, Dr, . . . . . . 179
Worms, expulsion of, . . 65, 102, 132, 133
germination of, . . . . . 102
Wound, . . . . . 155, 184
dressing and healing of, . . . 103, 180, 182
kinds of, . . . . . 155
*Wrightia antidysenterica*, . . . 104, 129

X

*Xanthoxylum rhetsa*, . . . . . 58

Y

Yajna, son of Ruchi, . . . . 28
Yajnavalkya, a law-giver, . . . . 16
Yajur Ved, . . . . 18, 19, 21, 22, 24
Yama, a demon, . . . . . 159
Yantra, apparatus, Plates I.-VI., . . 143, 146
surgical appliance, Plates VII.-VIII., . 172, 182, 188
Yantra-mukta, machine projectile, . . . 16
Yapya, diseases controllable by remedies, . . 164
Yashada, zinc, . . . . . . 134
Yashadapushpa, oxide of zinc, . . . 15, 136

PAGES

Yava, a measure, . . . . . 149, 150
Yavakshara, carbonate of soda, . . . 136
Yavani, . . . . . . 53
Yavasa, *Alhagi maurorum*, . . . . 128
Yayati, a king, . . . . . 194
Yellow pyrites, . . . . . 135
Yoga, conjunction of stars, . . . . 76
contemplative philosophy, . . . 17, 76
Yoga-sutras, a work by Patanjala, . . . 76
Yogi, an ascetic, . . . . 117, 146, 147
Yuga, an age, . . . . . 19, 35
Yunani, Ionian, . . . . . 126

Z

Zeus, . . . . . . 190
Zinc, . . . . . . 134, 141
Bhasma, calcined, . . . . 135
carbonate of, . . . . . 135
oxide of, . . . . . 15, 136
used medicinally, . . . . 65, 134
*Zingiber officinale*, . . . . . 123